Ecological Economics

Stanislav E. Shmelev

Ecological Economics

Sustainability in Practice

Springer

Dr. Stanislav E. Shmelev
University of Oxford
Queen Elizabeth House
Flat 3, Banbury House
1 Staverton Road
Oxfordshire
Oxford OX2 6XH
UK
stanislav.shmelev@googlemail.com

ISBN 978-94-007-1971-2 e-ISBN 978-94-007-1972-9
DOI 10.1007/978-94-007-1972-9
Springer Dordrecht Heidelberg London New York

Library of Congress Control Number: 2011939209

Printed on acid-free paper

Springer is part of Springer Science+Business Media (www.springer.com)

Contents

List of Boxes

List of Figures

List of Tables

Introduction

This book is devoted to the M.Sc. and first year Ph.D. students reading for degrees in Environmental Change and Management, Sustainability, Ecological Economics, Environmental Management, Philosophy, Politics and Economics and taking part in similar programmes. It is aimed to provide an overview of a range of new methodological tools: environmentally extended input–output analysis, multicriteria decision aid, optimization, geographical information systems, life cycle assessment, material flows analysis and modern applications of these tools to the most pressing today's problems: assessment of sustainability, climate change and renewable energy, loss of biodiversity, global resource use and sustainable waste management, corporate sustainability and other relevant themes.

There have been textbooks published on Environmental and Resource Economics i.e. Turner, Pearce and Bateman (1994), Hanley and Shorgen (2001), Perman et al. (2003). All of them as well as this present text have their peculiarities: Perman et al. devote considerable attention to the environmentally extended input–output analysis covered in this volume, however do not cover the important field of multicriteria decision aid. Turner, Pearce and Bateman was a groundbreaking text at the time but is a little bit out of date at the moment, it also involves a strong emphasis on monetisation and cost-benefit analysis, which is not shared by the author of the present volume. Hanley and Shorgen (2001) is more focused on market instruments and less on the systems perspective.

Several strong textbooks on Ecological Economics have been issued in the past, i.e. Daly and Farley (2004), which comprises chapters on macroeconomic theory (IS-LM model) and new ways of assessing sustainability (ISEW) but doesn't cover such important applied areas as corporate sustainability, renewable energy or waste management and is more targeted at the US audience. Common and Stagl is probably the best available modern text in Ecological Economics however it is a bit too long for a semester course (592 pp) and although the text covers very relevant areas of environmental policies, the environmental effects of international trade, and involves two applied chapters on climate change and biodiversity loss, it does not discuss such methodological tools as multi-criteria decision aid or explore applications of principles of sustainability in the urban or corporate context, and is written at a more elementary level, than e.g. Perman et al. (2003). Faber and Proops (1998) is a wonderful theoretical introduction to the field, the book has a distinct philosophical focus but does not have many practical applications and is a little bit out of date over 10 years after its publication.

The current text is designed to be a concise, crisp, and elegant guide packed with references for students with some background in economics, environmental science or mathematics aimed at developing their analytical skills required for redirecting our development path towards sustainability in government, international organisations, academia, non-profit sector and business. It builds on the idea that a significant adjustment of the current economic theories is required, which was recently supported by the emerged world economic crisis, the climatic and biodiversity crisis the world is currently facing and the enormously slow progress that has been made in the field of reorientation of the global economy towards sustainability. We have chosen a positive approach for problem solving and strategic development, which is aimed at educating the future decision makers and business leaders.

The content of the book is envisaged to be the following: the first part of the book is theoretical, it is designed to give the methodological background and the tools for subsequent analysis; the second part is devoted to the applications.

Chapter 1 presents the subject of ecological economics, the interaction between the economic system and the environment; Chap. 2 explores the ideas of material and energy flows from the point of view of industrial ecology; Chap. 3 explores the ethical and world systems basis for sustainability thinking; Chap. 4 looks at decision making and the methods that could be used to support such processes, especially Multicriteria Decision Aid; Chap. 5 studies the concept of externalities and macroeconomic basis for environmental policy; Chap. 6 explores the potential of environmentally extended input–output modelling for sustainability analysis; Chap. 7 looks at another important aspect of ecological economic analysis: macro assessment of sustainability, the method invented by the author of this book and essentially the application of multicriteria decision aid to the dynamic comparison of periods in a performance of a country or a region.

Part two includes a chapter on the renewable energy, biodiversity assessment, sustainable cities, regional waste management, and Corporate Sustainability. The author felt that such a composition of subjects will give the students a holistic perspective on sustainability issues.

I would like to express my sincere gratitude to Prof. David Orr for giving me the International Society for Ecological Economics membership as my 21st birthday present, my Ph.D. advisors Dr. Gerald Shalabin and Dr. John Powell, my parents, as well as Prof. John Proops, Prof. Beat Bürgenmeier, Prof. Robert Ayres, Prof. Jeroen van den Bergh, Prof. Peter Söderbaum, Prof. Joan Martinez-Alier and Prof. Bernard Roy for our discussions with them, their support and encouragement. I am particularly grateful to Dr. Barbara Cowell for carefully reading the manuscript and suggesting ways to improve the style. Chapter on sustainable cities is written in collaboration with Prof. Irina Shmeleva, chapter on sustainable waste management with Dr. John Powell.

I sincerely hope that the methods and ideas presented in this book are going to be taken on by the students and developed further by the next generation of economists. The students using this textbook will undoubtedly benefit from reading the original scientific papers quoted in the literature reviews in respected chapters. I would highly encourage the interested readers to find and explore the original sources. Each chapter in this book is designed in such a way that it could be read independently. All chapters taken together will give the reader a "bigger picture", an interdisciplinary and holistic perspective on ecological economics and sustainability analysis as seen by the author.

Oxford

Part I
Theory of Ecological Economics

Chapter 1
The Economic System and the Environment

Abstract The first chapter defines ecological economics as an interdisciplinary field of research focused on the interactions between the economy and the environment. Major milestones in the history of ecological economics are identified. Definitions from the founding fathers of ecological economics are given and key differences between the methods of environmental and ecological economics are explored. A conceptual graphical model of the economic system as seen by ecological economics is constructed. The model includes renewable and non-renewable resources, the recycling sector as well as major elements of the environmental system being affected by the economy. The chapter presents an overview of the applications of major ecological-economic methods to key ecological-economic problems according to the Scopus academic citation system. Major gaps in the literature are identified.

Keywords Ecological economics • Sustainability • Economic system • Problems • Methods

Definitions

When he (she) starts to study ecological economics, the student embarks on an exciting interdisciplinary journey, which will bring answers to important questions, help to understand the ecological-economic system in all its intricacy and lead to new insights. Ecological economics emerged as a response to the pressing environmental problems of the twentieth century and the inability of neoclassical economic theory to solve them or provide adequate explanations for the unprecedented decline in biodiversity, the changing climate, increased generation of waste, all caused by the pursuit of economic growth.

S.E. Shmelev, *Ecological Economics: Sustainability in Practice*,
DOI 10.1007/978-94-007-1972-9_1,

It is very natural to start such a journey with definitions by the pioneers:

Robert Costanza (Costanza 1989) defines the new interdisciplinary science in the following way:

- Ecological Economics addresses the relationships between ecosystems and economic systems in the broadest sense.

This very inclusive definition implies that the works on the Limits to Growth (Meadows and Club of Rome 1972), the first environmentally focused input–output studies (Leontief 1970), the first conceptual models where different material resources were considered as important inputs to industrial processes (Ayres et al. 1970) all belong to the interdisciplinary field of ecological economics.

John Proops (1989) suggested a more detailed and elaborate definition, differentiating (i) the scientific aims and problems and (ii) political and ethical issues:

Scientific Aims and Problems

- Establishing a historical perspective on social-natural interactions
- Finding a common language and a set of concepts for the analysis of economies and ecosystems
- The area of intersection between natural science and social science

Political and Ethical Issues

- As a forum and structuring for policy analysis
- A framework for the ethical analysis of intertemporal and interspecies choice
- The influencing of decision makers

According to this definition, such works as (Fischer-Kowalski 1998, Fischer-Kowalski and Hattler 1998) focusing on the historical dimension of the human appropriation of natural resources, the interdisciplinary works on the ability of systems to return back to undisturbed states, which is also called "resilience" (Holling 1973) and the works on means of taking nature into account when making decisions (Foster 1997) all form the first pillar of ecological economics according to Proops. Interestingly, Proops emphasises the second, transformative and interactive dimension of ecological economics, which is designed to be the policy forum for influencing decision makers.

Jeroen van den Bergh (2000) explicitly mentions all the constituent disciplines that interact to support ecological economics:

- EE integrates elements of economics, ecology, thermodynamics, ethics, and a range of other natural and social sciences to provide an integrated and biophysical perspective on environment-economy interactions, aimed at contributing to structural solutions to environmental problems.

This definition corresponds to the spirit of interdisciplinary works on the biosphere (Vernadsky 1929), shallow and deep ecology (Naess 1973), new ethical economics (Schumacher 1973), and sustainable cities (Shmelev and Shmeleva 2009).

Ecological and Environmental Economics

Ecological economics has been critical of the mainstream for failing to educate future decision makers in the spirit of socially inclusive, environmentally sound and economically sustainable development. Graduates of neoclassical economic departments continue to reproduce the logical errors of the theory in the real world, suggesting that unlimited economic growth will cure all the problems of modern civilization, that one can simply export waste to less developed countries and that one only needs to take into account the economic costs and benefits of climate change and biodiversity to make a decision on what to do and where to invest to tackle the problem. And although there were significant figures in the neoclassical era, who brought the pure economic analysis to new heights, such as Alan Marshall, the twentieth century brought with it new challenges and required new methods to address them.

If one opens an introductory neoclassical textbook of economics one is most likely to see a diagram similar to the one depicted in Fig. 1.1. It usually includes such agents as households, firms, government and foreign agents. In very advanced textbooks this diagram will have a box called "Nature" or ecosystems, with the flows of materials and energy emerging from it. The problem with this formulation is that the role of the environmental system as a support system for all processes carried out in the national economy (agricultural production, mining, deposition of waste) is not represented accurately and the environment is considered as a subsystem of the economy. Hence the attempts to apply economic valuation to environmental phenomena, which constitute a logical error.

The vision of the world, which we can see in diagrams similar to Fig. 1.1 can be explained by the historical inheritance. In the nineteenth and even the beginning of the twentieth century, the world economy was operating in what Herman Daly (2000) called the "empty world", depicted in Fig. 1.2.

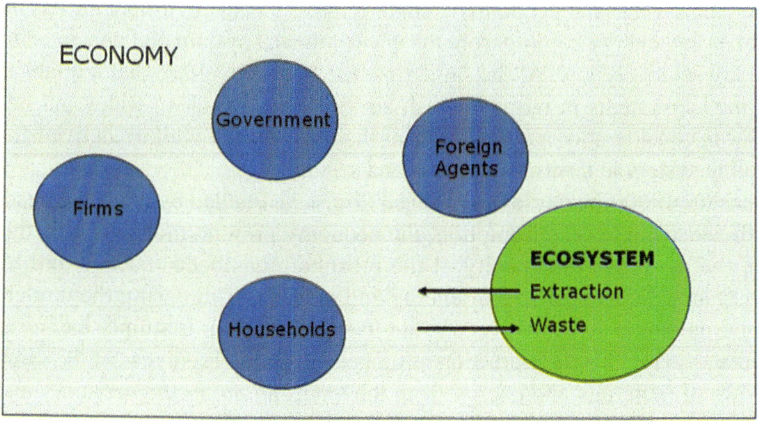

Fig. 1.1 Neoclassical view of the world

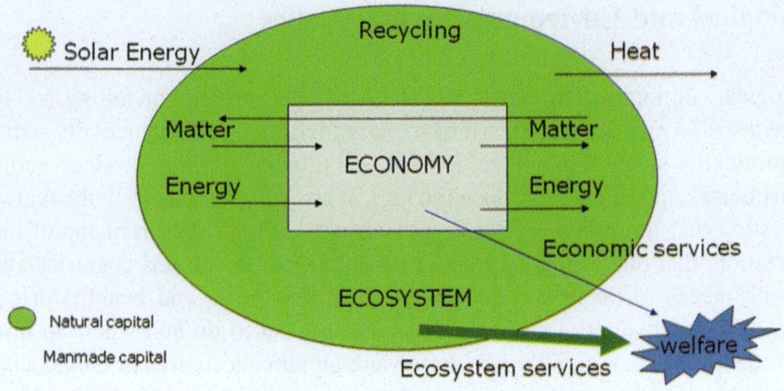

Fig. 1.2 "Empty" world, nineteenth and beginning of twentieth century

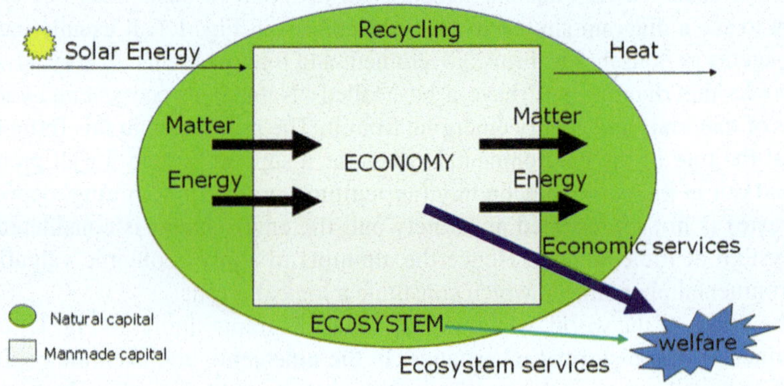

Fig. 1.3 "Full" world, 1960s onwards

We can see that the economy is small relative to the environment; the flows of resources and energy coming from the environmental system and deposited there as waste are relatively low. At the same time the flow of welfare that humans receive from the ecosystems in terms of fresh air, forest walks, clean water and beautiful scenery is considerably more significant than the flow of welfare derived from the economic system in terms of products and services.

Since the 1960s the world has changed (Fig. 1.3). Fuelled by the idea of economic growth and increased consumption, the economy grew tremendously to the point where the assimilative capacity of the biosphere has been reached and humans use very significant amounts of energy and materials, hindering the tendency for the environmental system to regenerate itself. Very often it simply has no time to regenerate, so rapid is the extraction of timber and other resources. At the same time, the flows of materials and energy from the environment to the economy and back to the environment become much more pronounced and the humans receive more welfare from the stream of goods and services (TV sets, mobile phones, cars, etc.) than from the environment. The overexploitation of the natural world has led to

Table 1.1 Differences between ecological and environmental economics (Source: van den Bergh 2000)

Ecological economics	Environmental and resource economics
1. Optimal scale	1. Optimal allocation and externalities
2. Priority to sustainability	2. Priority to efficiency
3. Needs fulfilled and equitable distribution	3. Optimal welfare to Pareto efficiency
4. Sustainable development, globally and North/South	4. Sustainable growth in abstract models
5. Growth pessimism and difficult choices	5. Growth optimism and "win-win" options
6. Unpredictable co-evolution	6. Deterministic optimisation of intertemporal welfare
7. Long-term focus	7. Short to medium term focus
8. Complete, integrative and descriptive	8. Partial, monodisciplinary and analytical
9. Concrete and specific	9. Abstract and general
10. Physical and biological indicators	10. Monetary indicators
11. Systems analysis	11. External costs and economic valuation
12. Multidimensional evaluation	12. Cost-benefit analysis
13. Integrated models with cause-effect relationships	13. Applied general equilibrium models with external costs
14. Bounded individual rationality and uncertainty	14. Maximisation of utility and profit
15. Local communities	15. Global market and isolated individuals
16. Environmental ethics	16. Utilitarianism and functionalism

increased CO_2 emissions and climatic changes, destruction of ecosystems and biodiversity, which stabilise the climatic system as well as the excessive pollution of the environment with waste, which is ever more apparent in the developing world.

The logic of ecological economics is that the world has changed tremendously and we need new conceptual tools to understand and manage the economic-environmental system. Ecological economists work across disciplines, building teams of experts and integrating knowledge to derive policy mechanisms, which help to prevent degradation and facilitate improvement.

By offering new methodological grounds, combining the advanced methods of environmentally extended input–output analysis, multicriteria decision aid, insights from ecology, biology, psychology and sociology, ecological economics aims to improve our understanding of sustainability and help to steer our economies in that direction.

Often there is confusion about the differences between ecological economics and the popular 1970s and 1980s school of environmental economics. Although the main focus of the two disciplines is similar, and one understands the value of the concept of externality and much of the analysis of environmental policy tools such as environmental taxes and their applications, which was prominent within environmental economics community, ecological economics is closer to the heart of the author for a number of reasons. Table 1.1 highlights the main differences between ecological and environmental economics as presented by Jeroen van den Bergh (2000).

Even if we focus here only on the most important ones, the differences will still be considerable. First of all, there is an explicitly long-term focus in ecological economics; the author would argue that it has multiple time scales, but because sustainability is a dynamic long-term phenomenon, we need to concentrate on the long term issues of resource use, accumulation of emissions, technological transformations, and evolutionary perspectives. The prioritising of sustainability in ecological economics corresponds to the prioritising of efficiency in environmental economics. The meeting of needs and equitable distribution in ecological economics is opposed to the optimal welfare and Pareto efficiency in environmental economics. Ecological economics focuses on physical and biological indicators whereas environmental economics emphasises monetary measures. The principle of multicriteria evaluation of ecological economics contrasts with the idea of cost-benefit analysis in environmental economics. The environmental ethics of ecological economics is a response to the utilitarianism and functionalism of environmental economics.

Systemic Vision

If we look at the more adequate descriptions of the interactions between the economic system and the wider environment depicted in Figs. 1.4–1.6 we notice significant differences from the worldview of the neoclassical approach, summarised by Fig. 1.1.

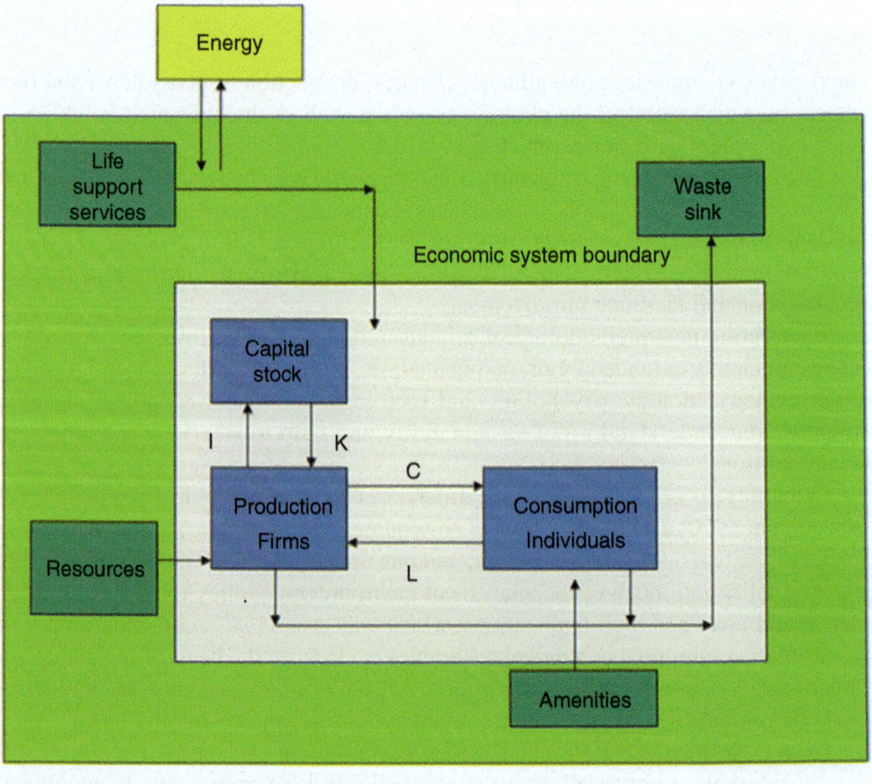

Fig. 1.4 Economic and environmental system (Adapted from Common and Stagl 2005)

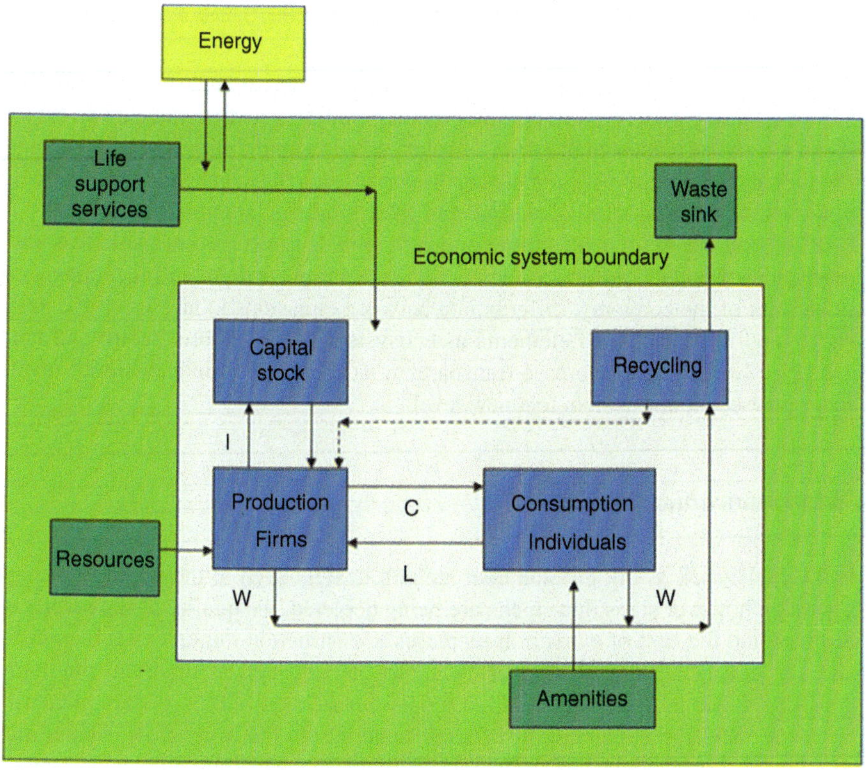

Fig. 1.5 Economic and environmental system: more realism

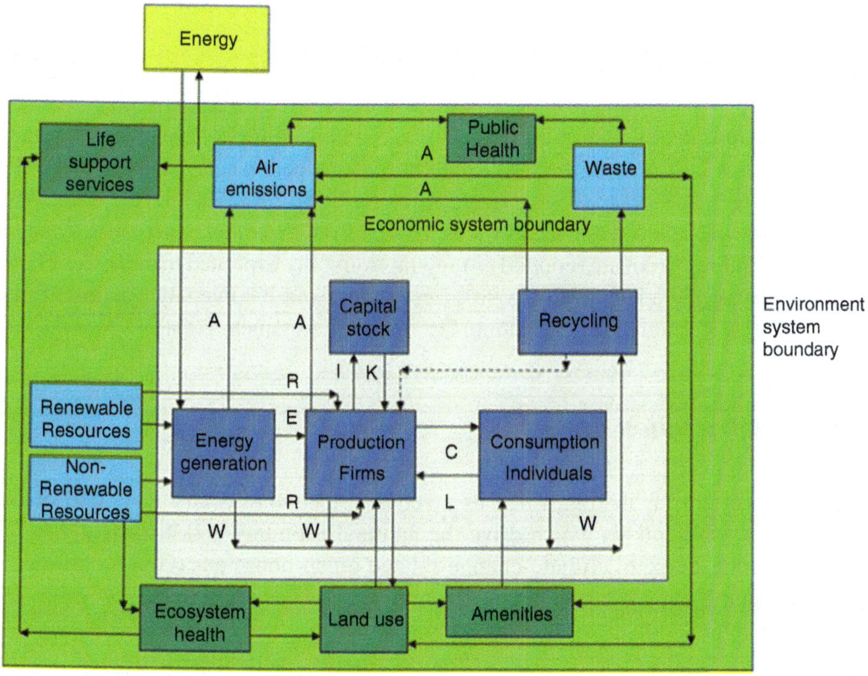

Fig. 1.6 Ecological-economic system: a realistic view

If we add the crucial elements of environmental resources and waste sinks, which the economic system uses all the time, specify energy as an external element to the system, arriving mostly in the form of solar radiation, and differentiate life support services and amenities, we can make an important step towards better understanding of the relationship between economic and environmental systems (Fig. 1.5).

If we complicate matters further and differentiate between renewable and non-renewable resources, introduce the energy generation module, add recycling as a subsystem of the economy, differentiate between emissions to air, water and solid waste, and introduce such elements as ecosystem health, public health and land use (Fig. 1.6), we will increase transparency and bring our understanding of the ecological-economic system to a new level.

Non-renewable Resources

Fossil fuels, such as oil, gas and coal are still heavily used in the world economy. There are however signs that stocks are being depleted, the quality of the resources declines and the cost of extraction increases. Construction minerals, such as stone and clay are in relative abundance, whereas metals, such as uranium, platinum, gold, copper, aluminium are in limited supply but are required for modern industrial processes; and the consumption of such materials is likely to grow. Sometimes shortage of one particular type of metal may result in the halting of the whole industrial process. The processes for the mining and production of many metals require substantial use of water and energy as well as resulting in considerable emissions.

Renewable Resources

Renewable resources such as forestry, fish stocks and other types of biomass, e.g. agricultural crops can provide harvests indefinitely if operated at sustainable levels. Unfortunately, fish stocks are being depleted, in the North-East Atlantic, the FAO (FAO 2009a) flags Atlantic cod and Haddock as ranging from exploited to depleted, Atlantic Salmon, Whiting, Trout are reported as ranging from fully exploited to depleted. Forest depletion is accelerating (FAO 2009b). Ecosystems and biodiversity continue to be depleted at an alarming rate (Millenium Ecosystem Assessment 2004).

Energy Generation

Energy generation is distinguished as a separate element because it is one of the most important elements which drive the national economy and historically it has been the main source of climate-change-related green-house gas (GHG) emissions. By redesigning our energy systems we could reduce our dependency on oil in the long-run, and link ourselves with the natural forces of wind, wave, and solar energy, thereby reducing the climate change impacts of our economic development.

Emissions and Waste

By explicitly considering emissions and waste we are exploring the issues of resource recycling, thereby saving energy and, working with a smaller resource stock, minimising resource use. Taking into account the effects of contaminating water and air with chemical pollutants will help us understand the side effects of economic development and especially its impact on the health of the public and the state of the ecosystems and of biodiversity.

Land Use

By focusing on land use we can explicitly take into account the spatial aspect of economic development in the spirit of classical economists, looking at the value and productivity of land as one of the main types of capital. Land-use change, urbanisation, agricultural expansion and ecosystem degradation would be particularly relevant here.

Early History

The early history and evolution of ecological economics as a discipline is very well captured by several important publications, among which is, of course, the first issue of the journal, Ecological Economics, published in 1989. The first publications introducing ecological economics published internationally were (Ayres et al. 1970, Boulding 1966, Georgescu-Roegen 1971, 1976, Leontief 1970, Proops 1983, 1989, Stanfield 1983); and most importantly (Christensen 1989, Common and Perrings 1992, Costanza 1989, 1996, Costanza and Daly 1987a, b, Friend 1996, Hourcade et al. 1992, Martinez-Alier 1987, McGlade 1990, Munda 1997, Norgaard 1989, Pearce 1987, Perrings 1986, 1995, Turner et al. 1995). I would argue that publications by Robert Ayres, Wassily Leontief, H. Odum, Robert Costanza, Herman Daly, Joan Martinez-Alier, Charles Perrings, John Proops and David Pearce were most influential in establishing the foundation of the new interdisciplinary field.

Later publications, including the 10 year anniversary article by Robert Costanza (Costanza and King 1999), as well as two articles by Inge Røpke on the history of ecological economics (Røpke 2004, 2005) complete the overview of the beginnings.

Key Dimensions

Ecological economics evolved along a series of dimensions, the most important of which were the limits of economic growth paradigm (Ayres 1998, Boulding 1966, Daly 1972, 1974a, b, 1977, 1987, 1990, 2000, Daly and Cobb 1989, Shmelev and Rodríguez-Labajos 2009); the idea of incommensurability of values and the use of multiple criteria methods in decision making (Martinez-Alier et al. 1998, Munda 1995, 2005a, b); democracy and institutional economics: (Söderbaum 1992, 1994, 1999, 2000, 2004); the use of energy in economic analysis: (Cleveland et al. 1984, Costanza 1980, Costanza and Herendeen 1984, Costanza and Neill 1984, Georgescu-Roegen 1971, Huettner and Costanza 1982). Equally prominent were the works on the analysis of interactions between the economic and environmental systems: (Ayres 1978, Ayres and Kneese 1969, Ayres and Simonis 1994, Ayres et al. 1970, Leontief 1970, 1977, Leontief and Ford 1972); ecosystem services: (Costanza 2008, Costanza and Mageau 1999, Costanza et al. 1998, 2007) and interdisciplinary works, which included many regional applications to the issues of water, energy, resource use, biodiversity, and waste management.

We can differentiate the ecological-economic problems the world is facing today on the basis of scale:

Global

- Climate Change
- Biodiversity Loss
- International Trade and the Environment
- Sustainable Water Management

National

- Sustainability at the Macro Scale
- Industrial Ecology
- Renewable Energy
- Sustainable Transport
- Responsible Consumption
- Land Use Change
- Ecosystem Health

Regional/local

- Sustainable Cities
- Green Business
- Waste management
- Sustainable Planning
- Eco Design

In order to analyse the often complex multi-stakeholder and multi-system problems we need to use a range of sophisticated methods, which have evolved over the course of the past 50 years. These methods include:

Key Methods

- Systems Analysis
- Environmental Accounting

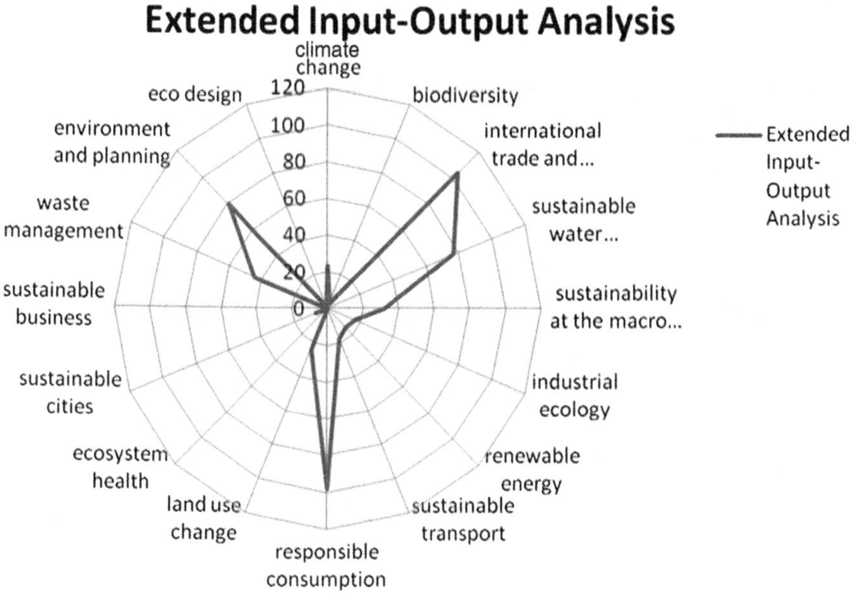

Fig. 1.7 Problems addressed with the help of environmentally extended input–output analysis (Source: Scopus)

- Environmentally Extended Input–output Analysis
- Energy Analysis
- Systems Dynamics
- Simulation Modelling
- Multicriteria Assessment
- Agent-Based Modelling
- Material Flows Analysis
- Life Cycle Analysis
- Environmental Valuation
- Optimization
- Ecosystem Services Analysis
- Evolutionary Analysis
- Stakeholder Analysis
- Quality of Life Analysis
- Citizen's Jury

These lists may be incomplete, but they give the reader an adequate view of the methods of ecological-economic analysis which are applied today at the cutting edge of the sustainability science.

Figures 1.7–1.11 offer snapshots of the problems and the methods by which they are usually tackled in ecological-economic analysis. We can see that environmentally extended input–output analysis has been most frequently applied in order to address the issues of international trade and the environment, sustainable water use, responsible

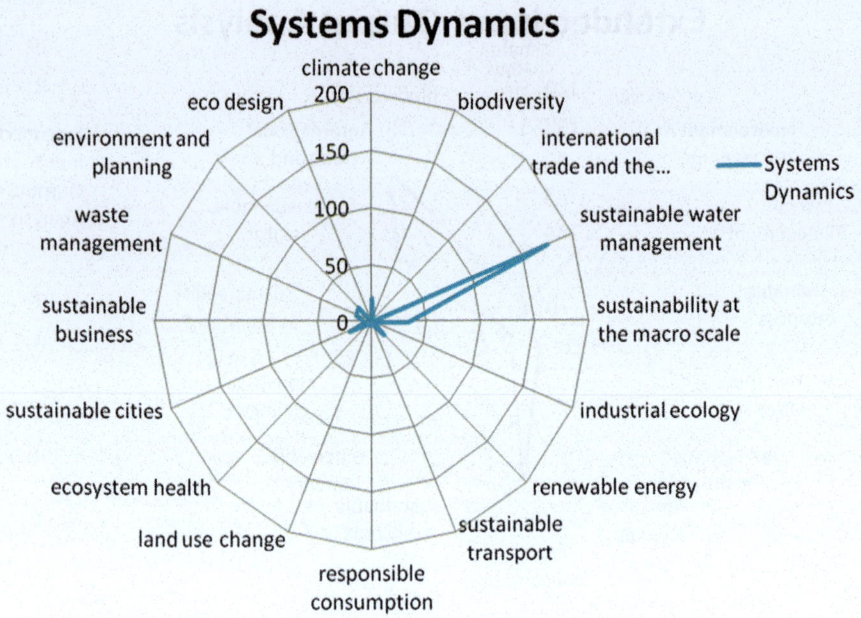

Fig. 1.8 Problems addressed with the help of systems dynamics approach (Source: Scopus)

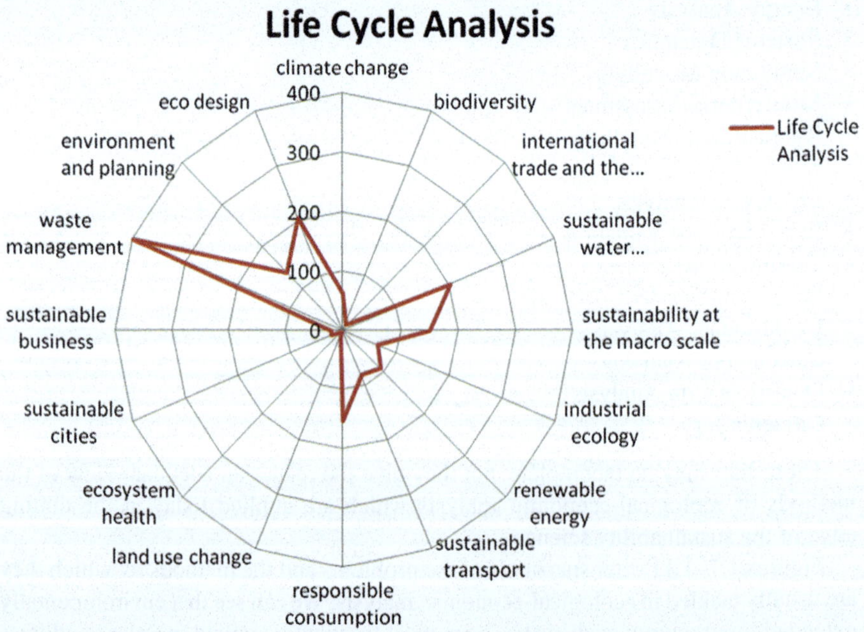

Fig. 1.9 Problems addressed with the help of life cycle analysis (Source: Scopus)

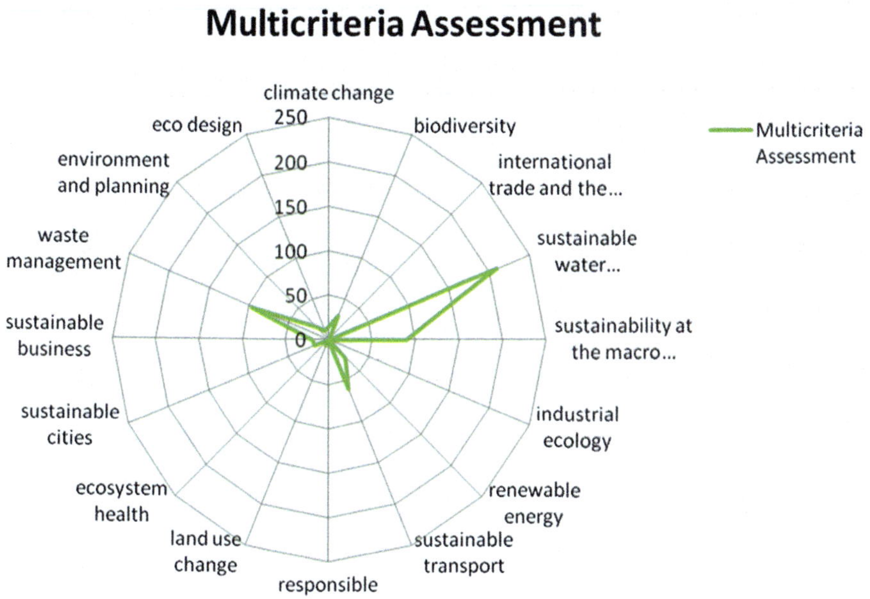

Fig. 1.10 Problems addressed with the help of multicriteria decision aid (Source: Scopus)

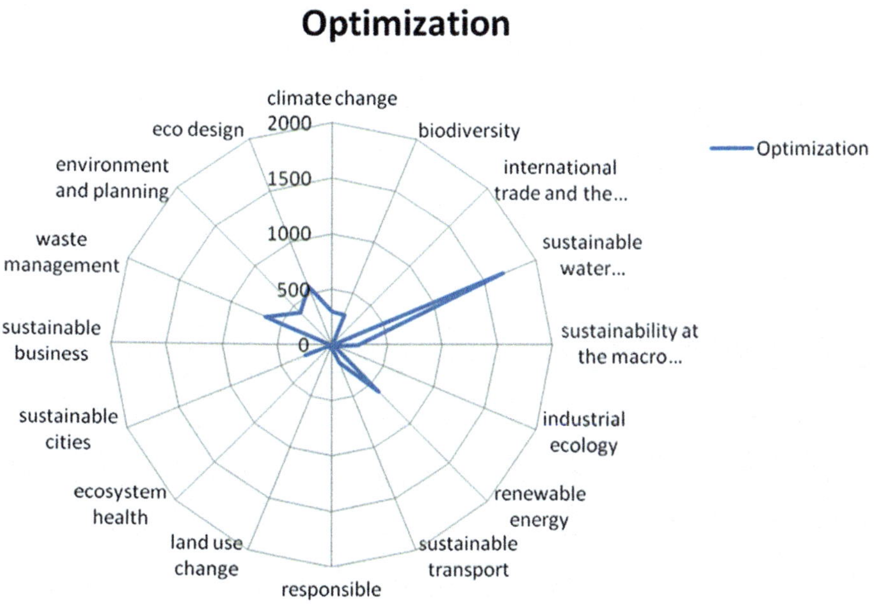

Fig. 1.11 Ecological-economic problems addressed with the help of optimization tools (Source: Scopus)

consumption, waste management and general issues of environment and planning (Fig. 1.8). The systems dynamics methodology has been most often used in sustainable water management (Fig. 1.8). Life cycle analysis has been frequently used in eco design, waste management, sustainable water management and responsible consumption (Fig. 1.9). Multicriteria decision aid is frequently used in the areas of water management, sustainable transport and biodiversity assessment (Fig. 1.10). Optimization is used in sustainable water management, eco-design, renewable energy, and waste management applications (Fig. 1.11). Such a diagrammatic "clustering" of the problem/method field can be very useful in identifying gaps in the literature and directions of further research.

References

Ayres RU (1978) *Resources, Environment and Economics. Applications of the Materials/Energy Balance Principle* New York, Wiley

Ayres RU (1998) *Turning Point: An End to the Growth Paradigm* London, Earthscan

Ayres RU and Kneese AV (1969) Production, Consumption, and Externalities *The American Economic Review* 59.3 282-297

Ayres RU and Simonis U (1994) *Industrial Metabolism: Restructuring for Sustainable Development* Tokyo, United Nations University Press

Ayres RU, D'Arge R and Kneese AV (1970) *Economics and the Environment: A Materials Balance Approach* Washington, Resources for the Future

Boulding K (1966) The Economics of the Coming Spaceship Earth

Christensen PP (1989) Historical Roots for Ecological Economics – Biophysical Versus Allocative Approaches *Ecological Economics* 1.1 17-36

Cleveland CJ et al (1984) Energy and the US Economy: A Biophysical Perspective *Science* 225.4665 890-897

Common M and Perrings C (1992) Towards an Ecological Economics of Sustainability *Ecological Economics* 6.1 7-34

Common M and Stagl S (2005) *Ecological Economics: An Introduction*, Cambridge University Press

Costanza R (1980) Embodied Energy and Economic Valuation *Science* 210.4475 1219-1224

Costanza R (1989) What is Ecological Economics? *Ecological Economics* 1.1 1-7

Costanza R (1996) Ecological Economics: Reintegrating the Study of Humans and Nature *Ecological Applications* 6.4 978-990

Costanza R (2008) Ecosystem Services: Multiple Classification Systems Are Needed *Biological Conservation* 141.2 350-352

Costanza ER and Daly HE (1987a) Ecological Economics *Ecological Modelling* 38.1–2 1-190

Costanza ER and Daly HE (1987b) Toward an Ecological Economics *Ecological Modelling* 38.1–2 1-7

Costanza R and Herendeen RA (1984) Embodied Energy and Economic Value in the United States Economy: 1963, 1967 and 1972 *Resources and Energy* 6.2 129-163

Costanza R and King J (1999) The First Decade of Ecological Economics *Ecological Economics* 28.1 1-9

Costanza R and Mageau M (1999) What Is a Healthy Ecosystem? *Aquatic Ecology* 33.1 105-115

Costanza R and Neill C (1984) Energy Intensities, Interdependence, and Value in Ecological Systems: A Linear Programming Approach *Journal of Theoretical Biology* 106.1 41-57

Costanza R et al (1998) The Value of Ecosystem Services: Putting the Issues in Perspective *Ecological Economics* 25.1 67-72

Costanza R et al (2007) Biodiversity and Ecosystem Services: A Multi-scale Empirical Study of the Relationship Between Species Richness and Net Primary Production *Ecological Economics* 61.2–3 478-491

Daly HE (1972) In Defense of a Steady-State Economy American *Journal of Agricultural Economics* 54.5 945-954

Daly HE (1974a) Steady-State Economics Versus Growthmania: A Critique of the Orthodox Conceptions of Growth, Wants, Scarcity, and Efficiency *Policy Sciences* 5.2 149-167

Daly HE (1974b) The Economics of the Steady State *The American Economic Review* 64.2 15-21

Daly HE (1977) *Steady State Economics: The Economics of Biophysical Equilibrium and Moral Growth* San Francisco, W.H. Freeman

Daly HE (1987) The Economic Growth Debate: What Some Economists Have Learned but Many Have Not *Journal of Environmental Economics and Management* 14.4 323-336

Daly HE (1990) Toward Some Operational Principles of Sustainable Development *Ecological Economics* 2.1 1-6

Daly H (2000) *Ecological Economics and the Ecology of Economics: Essays in Criticism* Cheltenham, Edward Elgar

Daly HE and Cobb J (1989) *For the Common Good: Redirecting the Economy Toward Community, the Environment, and a Sustainable Future* Boston, Beacon Press

FAO (2009) The State of the World Fisheries and Aquaculture 2008 Available at ftp://ftp.fao.org/docrep/fao/011/i0250e/i0250e.pdf

FAO (2009) State of the Worlds Forests Available at ftp://ftp.fao.org/docrep/fao/011/i0350e/i0350e.pdf

Fischer-Kowalski M (1998) Society's Metabolism: The Intellectual History of Materials Flow Analysis, Part I, 1860–1970 *Journal of Industrial Ecology* 2.1 61-78

Fischer-Kowalski M and Hattler W (1998) Society's Metabolism: The Intellectual History of Materials Flow Analysis, Part II, 1970–1998 *Journal of Industrial Ecology* 2.4 107-136

Foster J (1997) *Valuing Nature? Ethics, Economics and the Environment* London, Routledge

Friend AM (1996) Sustainable Development Indicators: Exploring the Objective Function *Chemosphere* 33.9 1865-1887

Georgescu-Roegen N (1971) *The Entropy Law and the Economic Process* Cambridge, Harvard University Press

Georgescu-Roegen (1976) *Energy and Economic Myths. Institutional and Analytical Economic Essays* Elmsford, Pergamon Press

Holling C (1973) Resilience and Stability of Ecological Systems *Annual Review of Ecology and Systematics* 4.1 1-23

Hourcade J-C, Salles J-M and Théry D (1992) Ecological Economics and Scientific Controversies. Lessons from Some Recent Policy Making in the EEC *Ecological Economics* 6.3 211-233

Huettner DA and Costanza R (1982) Economic Values and Embodied Energy *Science* 216.4550 1141-1143

Leontief W (1970) Environmental Repercussions and the Economic Structure: An Input-Output Approach *The Review of Economics and Statistics* 52.3 262-271

Leontief W (1977) Natural Resources, Environmental Disruption, and the Future World Economy *Journal of International Affairs* 31.2 267

Leontief WW and Ford D (1972) Air Pollution and the Economic Structure: Empirical Results of Input-Output Computations *in* Brody A and Carter AP *Input-Output Techniques* Amsterdam, North-Holland Publishing Company

Martinez-Alier J (1987) *Ecological Economics: Energy, Environment and Society. Ecological Economics: Energy, Environment and Society* Available at http://www.scopus.com/inward/record.url?eid=2-s2.0-0023491881&partnerID=40

Martinez-Alier J, Munda G and O'Neill J (1998) Weak Comparability of Values as a Foundation for Ecological Economics *Ecological Economics* 26.3 277-286

McGlade J (1990) Ecological Economics *Trends in Ecology and Evolution* 5.12 396-397

Meadows D and Club of Rome (1972) *The Limits to Growth a Report for the Club of Rome's Project on the Predicament of Mankind* New York, Universe Books

Millenium Ecosystem Assessment (2004) *Ecosystems and Human Well-Being: Current State and Trends Volume 1* Washington DC, Island Press
Munda G (1995) *Multicriteria Evaluation in a Fuzzy Environment* Heidelberg, Physica-Verlag
Munda G (1997) Environmental Economics, Ecological Economics, and the Concept of Sustainable Development *Environmental Values* 6.2 213-233
Munda G (2005a) "Measuring Sustainability": A Multi-Criterion Framework *Environment Development and Sustainability* 7.1 117-134
Munda G (2005b) Multiple Criteria Decision Analysis and Sustainable Development *in Multiple-Criteria Decision Analysis. State of the Art Surveys* New York, Springer 953-986
Naess A (1973) Shallow and Deep, Long-Range Ecology Movement. A Summary. INQUIRY *An Interdisciplinary Journal of Philosophy* 16.1 95-100
Norgaard RB (1989) The Case for Methodological Pluralism *Ecological Economics* 1.1 37-57
Pearce D (1987) Foundations of an Ecological Economics *Ecological Modelling* 38.1–2 9-18
Perrings C (1986) Conservation of Mass and Instability in a Dynamic Economy-Environment System *Journal of Environmental Economics and Management* 13.3 199-211
Perrings C (1995) Ecology, Economics and Ecological Economics *Ambio* 24.1 60-64
Proops JLR (1983) Organisation and Dissipation in Economic Systems *Journal of Social and Biological Systems* 6.4 353-366
Proops JLR (1989) Ecological Economics: Rationale and Problem Areas *Ecological Economics* 1.1 59-76
Røpke I (2004) The Early History of Modern Ecological Economics *Ecological Economics* 50.3–4 293-314
Røpke I (2005) Trends in the Development of Ecological Economics From the Late 1980s to the Early 2000s *Ecological Economics* 55.2 262-290
Schumacher EF (1973) *Small Is Beautiful. Economics as if People Mattered* New York, Harper & Row
Shmelev SE and Rodríguez-Labajos B (2009) Dynamic Multidimensional Assessment of Sustainability at the Macro Level: The Case of Austria *Ecological Economics* 68.10 2560-2573
Shmelev SE and Shmeleva IA (2009) Sustainable Cities: Problems of Integrated Interdisciplinary Research *International Journal of Sustainable Development* 12.1/2009 4-23
Söderbaum P (1992) Neoclassical and Institutional Approaches to Development and the Environment *Ecological Economics* 5.2 127-144
Söderbaum P (1994) Actors, Ideology, Markets. Neoclassical and Institutional Perspectives on Environmental Policy *Ecological Economics* 10.1 47-60
Söderbaum P (1999) Values, Ideology and Politics in Ecological Economics *Ecological Economics* 28.2 161-170
Söderbaum P (2000) *Ecological Economics: A Political Economics Approach to Environment and Development* London, Earthscan
Söderbaum P (2004) Democracy, Markets and Sustainable Development: The European Union as an Example *European Environment: The Journal of European Environmental Policy (Wiley)* 14.6 342-355
Stanfield JR (1983) Toward an Ecological Economics *International Journal of Social Economics* 10.5 27-37
Turner RK, Perrings C and Folke C (1995) Ecological Economics: Paradigm or Perspective. Working Paper – Centre for Social & Economic Research on the Global Environment (GEC 95–17) Available at http://www.scopus.com/inward/record.url?eid=2-s2.0-0029484516&partnerID=40
van den Bergh JCJM (2000) Ecological Economics. Themes, Approaches, and Differences with Environmental Economics. Available at: http://www.tinbergen.nl/discussionpapers/00080.pdf
Vernadsky VI (1929) *La Biosphere* Paris, Librairie Félix Alcan

Chapter 2
Industrial Ecology: Material and Energy Flows, Life Cycle Analysis

Abstract This chapter is devoted to the conceptual foundations of industrial ecology, an interdisciplinary field that draws parallels between the natural world of organisms, their use of energy and resources and the interactions between them and the world of enterprises that also interact, use energy and resources and differ from natural entities in peculiar ways. The subject of biogeochemical cycles, underpinning the ideas of industrial ecology is introduced alongside three major analytical methods which can be used to study the environmental effects of economic interactions: those of life cycle analysis, material flows analysis and environmentally extended input–output analysis. A series of diagrams illustrates the global distribution of material flows of a particular kind or the differences between industrial and ecological systems.

Keywords Industrial ecology • Life cycle assessment • Material flows analysis • Environmentally extended input–output analysis • Biogeochemical cycles

Biogeochemical Cycles

The problem of biogeochemical cycles was first addressed by Vladimir Vernadsky (1924, 1926, 1929, 1940). His discoveries later formed the basis for Earth Systems Science. The most important cycles that have been studied in great detail are those of Carbon (C), Nitrogen (N), Phosphorus (P), Sulphur (S), described extensively in (Ayres 2002).

The understanding of the Earth system as a complex self-regulating, non-linear entity composed of various subsystems leads to other important works in the field: (Ayres 1978, Ayres et al. 1970, Lovelock 1972, Margulis and Lovelock 1978, Moiseev 1978, Moiseev et al. 1983, 1985). Much success in interdisciplinary research into what was later to become global sustainability can be attributed to the special role of the International Institute for the Applied Systems Analysis (IIASA) located in Laxenburg, Austria. The institute brought together scientists from across the globe to collaborate on Earth Systems Science and complexity.

S.E. Shmelev, *Ecological Economics: Sustainability in Practice*,
DOI 10.1007/978-94-007-1972-9_2, © Springer Science+Business Media B.V. 2012

Industrial Ecology

The concept of industrial ecology emerged in several places independently, which is excellently described in two historical overviews of the development of this field (Fischer-Kowalski 1998, Fischer-Kowalski and Hattler 1998). The idea of industrial ecology was first proposed by Watanabe in a project, devoted to the study of resource dependency in the Japanese economy (Duchin and Hertwich 2003), and a little later Robert Ayres independently developed the principles of this emerging discipline (Ayres 1978, Ayres and Ayres 2002, Ayres and Simonis 1994). The latter has been one of the true pioneers in the field of the analysis of economy-environment interactions: a formal mathematical framework for tracing residual flows in the economy was offered in (Ayres and Kneese 1969), ideas of a stationary state economy were explored in (Ayres and Kneese 1971), the ideas on the interaction between the economy and the environment resulted in a fundamental treatise (Ayres et al. 1970). These ideas were clearly influenced by the work of Wassily Leontief in the field of input–output analysis in the USA economy (Leontief 1936, 1949, 1952), and more especially by the environmentally extended applications of the input–output analysis which appear in (Leontief 1970, 1974, 1977a). Leontief built a conceptual link between the structure of the economy and the interdependent economic sectors and the environmental impacts of economic activity, namely air pollution.

Industrial ecology draws parallels between the ecological webs of the natural world and the economic webs of corporations and consumers (Table 2.1).

In the comparison between the biological and industrial organisms from the point of view of their resource and energy transformation the following conclusions are

Table 2.1 Biological and industrial organisms (Source: Graedel and Allenby 2002)

Biological organisms	Industrial organisms
Are able to act independently (differ in their degree of independence)	**Independent actors** (use and transform resources)
Use energy and material resources (transform resources into new forms, suitable for use, generate heat from the rest, and release residues).	**Yes** (residues of energy and materials are emitted into the environment)
Are able to reproduce themselves (life time and numbers of offspring vary)	**Reproduction is not the purpose** (creation of a product is). They do reproduce, however it is a function of specialised external agents
React to external impacts (temperature, humidity, availability of resources, potential partners for reproduction)	**Yes** (react to availability of resources, potential clients, prices, etc.)
All multicellular organisms are developing from a single cell and pass through several stages of development	**Not really.** Most plants and companies do evolve, but they do not follow a systematic and predictable sequence of life stages of biological organisms
Have a fixed life time	**Yes**, this property can be observed

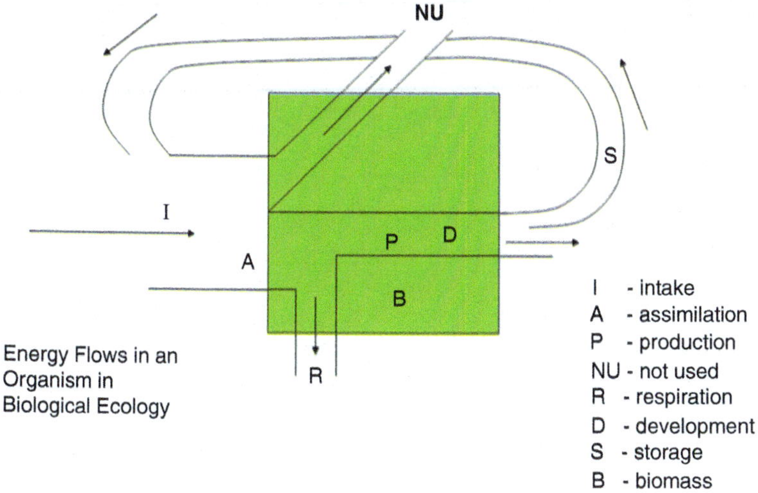

Fig. 2.1 Energy flows in a biological organism

drawn (Graedel and Allenby 2002): both can act independently, but differ in their degree of independence; both are using energy and materials and emit residues into the environment; reproduction is a unique property of biological organisms, which is not the purpose of industrial organisms, for which the creation of the product is a goal; they both react to external influences (temperature, humidity and the availability of resources, potential partners for reproduction for biological organisms and availability of resources, prices, potential clients for industrial organisms); biological organisms pass through several stages of development, which to some extent can be applicable at the industrial level (start-ups, industry pioneers, multinationals) and they both have a fixed lifetime.

On the other hand, the energy flows passing through both types of organisms differ in some way (Figs. 2.1 and 2.2).

The energy flows coming through the food-chain in the ecological and industrial systems also differ (Figs. 2.3 and 2.4).

The differences are largely due to the way energy is used and transformed (Figs. 2.3 and 2.4). Industrial ecosystems by contrast with ecological systems require external energy at every stage of the production process from primary producer to tertiary consumer, whereas in ecological systems, the upper levels of the food chain consume energy embodied in the organisms of the lower levels.

Three major methods are used to deal with the physical interactions between the economy and the environment. They are: life cycle analysis, focused at the level of product, a production line or a region; material flows analysis, usually focused at the level of the national economy or a region and the environmentally extended input–output analysis, which provides the connection between physical interactions and economic interdependencies. We will consider each of these methods in turn.

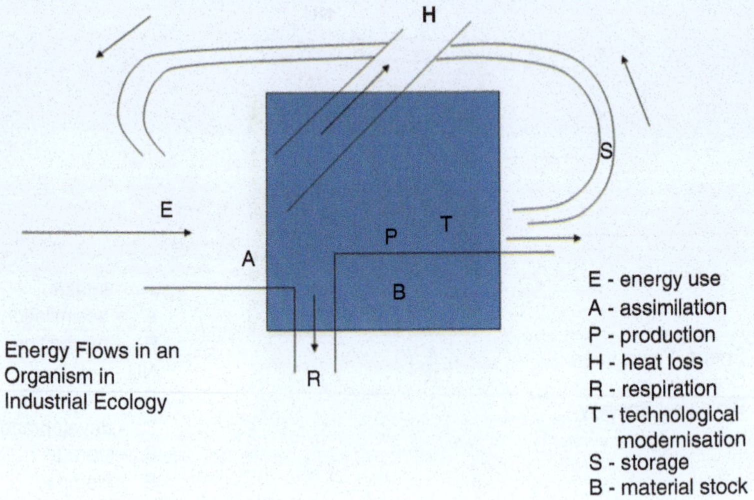

Fig. 2.2 Energy flows in an industrial organism (Adapted from Graedel and Allenby 2002)

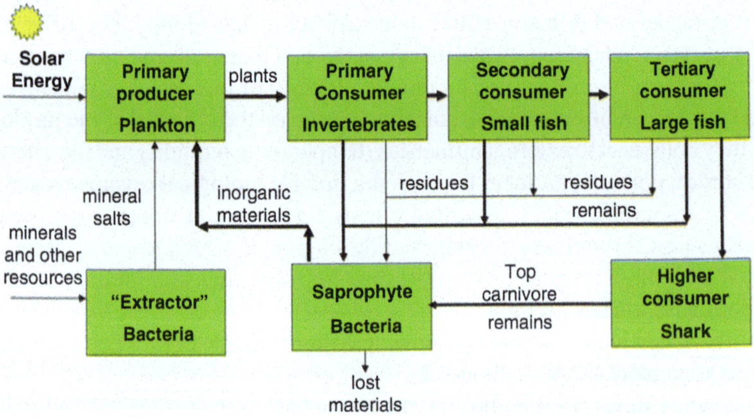

Fig. 2.3 Biological food chain (Sea) (Adapted from Graedel and Allenby 2002)

Life Cycle Analysis

Life cycle analysis or life cycle assessment was introduced by the International Organization for Standardization within its 14,000 set of environmental management standards (International Organization for Standardization 2007):

Life Cycle Assessment (LCA), Fig. 2.5, is a technique for assessing the potential environmental aspects and potential aspects associated with a product (or service), by:

- compiling an inventory of relevant inputs and outputs,
- evaluating the potential environmental impacts associated with those inputs and outputs,

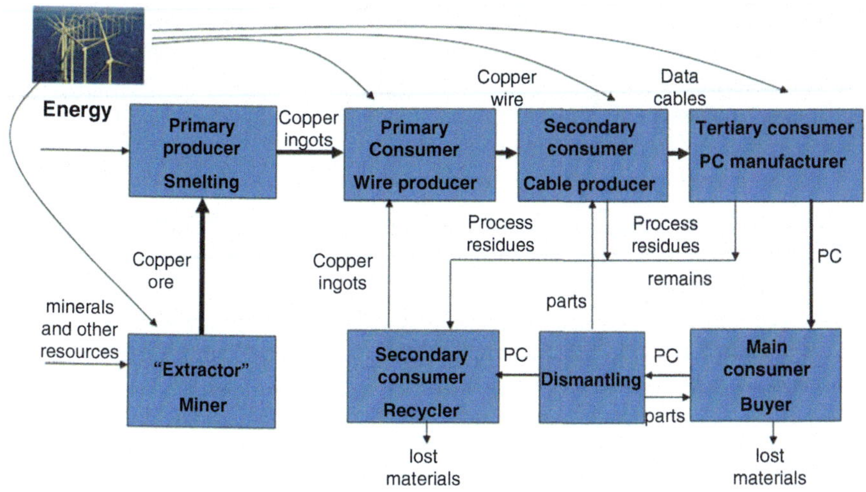

Fig. 2.4 Industrial food chain (Adapted from Graedel and Allenby 2002)

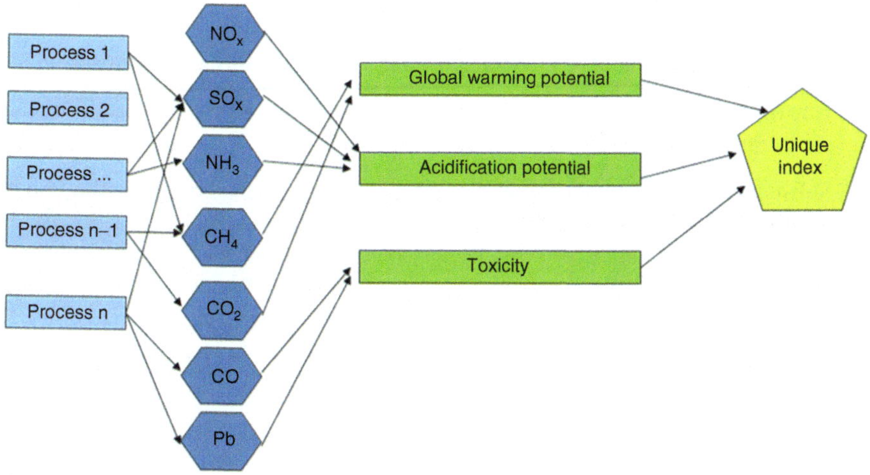

Fig. 2.5 Life cycle analysis, the process flow diagram

- interpreting the results of the inventory and impact phases in relation to the objectives of the study.

The analysis usually consists of four main stages:

1. **Initial phase**: setting the system boundaries, defining the problem and establishing an inventory of important parameters.
2. **Inventory Phase**: a detailed description of raw materials and energy inputs used at all points and emissions, effluent and solid waste outputs.

Examples of output are resource depletion (e.g. material and energy), pollutant emissions and discharges of chemical or physical load (e.g. substances, heat, and noise).

3. **Impact Assessment Phase**: relating the identified inputs and outputs to the environmental impacts (often called Life Cycle Impact Assessment). It involves the following components (the first three are mandatory, the others optional):
   - Selection of impact categories, category indicators and characterization models. Impact categories are selected and defined with respect to the goal and scope of the LCA.
   - Assignment of LCI results (Classification). The environmental loads are classified according to the impact categories. (Some environmental loads belong to more than one impact category.)
   - Calculation of category indicator results (Characterization). The category indicator is modelled for the different environmental loads which cause environmental impacts e.g. the Global Warming Potential.
   - Normalisation. Expressing category indicators relative to a standard e.g. tonne of CO_2 equivalent.
   - Grouping. Sorting and possibly ranking of impact categories.
   - Weighting. Expressing the (subjective) importance of an impact category: often the categories are sorted by theme or damage category.
   - Data Quality Analysis. Understanding the reliability of the indicator results.
4. **Improvement Phase**: using information obtained in analysis to improve overall environmental performance.

Substantial ecological economic literature has been devoted to the methodology and applications of life cycle analysis: (Ayres 1995, Ayres and Martinas 1992, Ayres et al. 1998, Azapagic and Clift 1999, Bengtsson 2001, Bouman et al. 2000, Boustead 1993, Carlson et al. 1998, De Udo Haes 1999, Fava 1997, Guinée and Heijungs 1993, Guinée et al. 1993, Haes et al. 2004, Hanssen and Asbjørnsen 1996, Heijungs and Guinée 1993, Heijungs and Suh 2002, 2006, Reinout Heijungs and United Nations Environment Programme 1996, Tukker et al. 1997).

Material Flows Analysis

Following the organisation of the United Nations System of National Accounts (United Nations 1947, 1953, 1968, 1993b, 2009) and the research started in the early works by Robert Ayres and colleagues in the USA (Ayres 1978, Ayres and Simonis 1994), Ayres et al. 1970, Konstantin Gofman in USSR (Gofman 2007) and Ernst von Weizacker in Germany (Weizsäcker and Club of Rome 1998) material flows analysis took shape over the course of the past 40 years and was formalised in an United Nations System of Environmental and Economic Accounting (United Nations 1993a, 2003) and later in the European Environmental Agency document on MFA (EEA 1999).

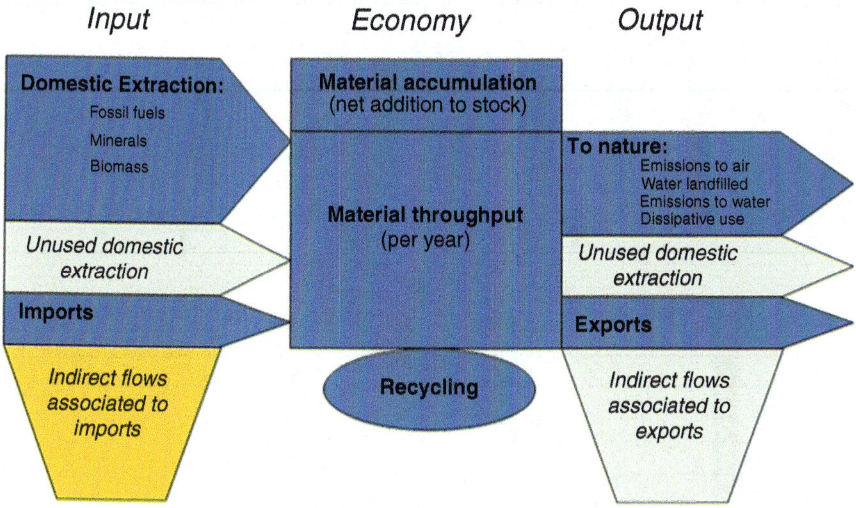

Fig. 2.6 Material flows analysis conceptual framework

The main features of the method (Fig. 2.6) are the aggregate approach to resource use accounting and the differentiation between domestic and imported/exported flows. An additional feature, prominent in the material flows analysis is usually referred to as indirect flows or unused fraction, the by-product of the mining and quarrying activity. The material throughput and material accumulation (net addition to stock) are complemented by recycling, aimed at a reduction in material throughput and energy use and an increase in the resource efficiency of the economy.

The author of this book was involved in 2003–2004 in the development of the Global Material Flows Database, which, following the European guidelines expressed in (EEA 1999), comprised an eight-level classification of materials extending to 400 positions at the 8th level for all countries of the world for the period 1980–2003 (Fig. 2.7).

A good example of the power of the global Material Flows Database might be the GIS based diagram shown in Fig. 2.8. The diagram depicts the world domestic extraction of the biomass item, Blueberries. Such an exotic example nevertheless allows us to assess the extraction of a very specific type of biomass across the globe and find the leading producing nations: Canada, USA, Poland, Romania, the Netherlands, etc.

While the first two levels of classification are represented by a natural system of inputs, outputs and net addition to stock (Table 2.2), the methodology for the output side of the method has only just been finalised in the European Union more than 10 years after the publication of the input accounting methodologies (Table 2.3).

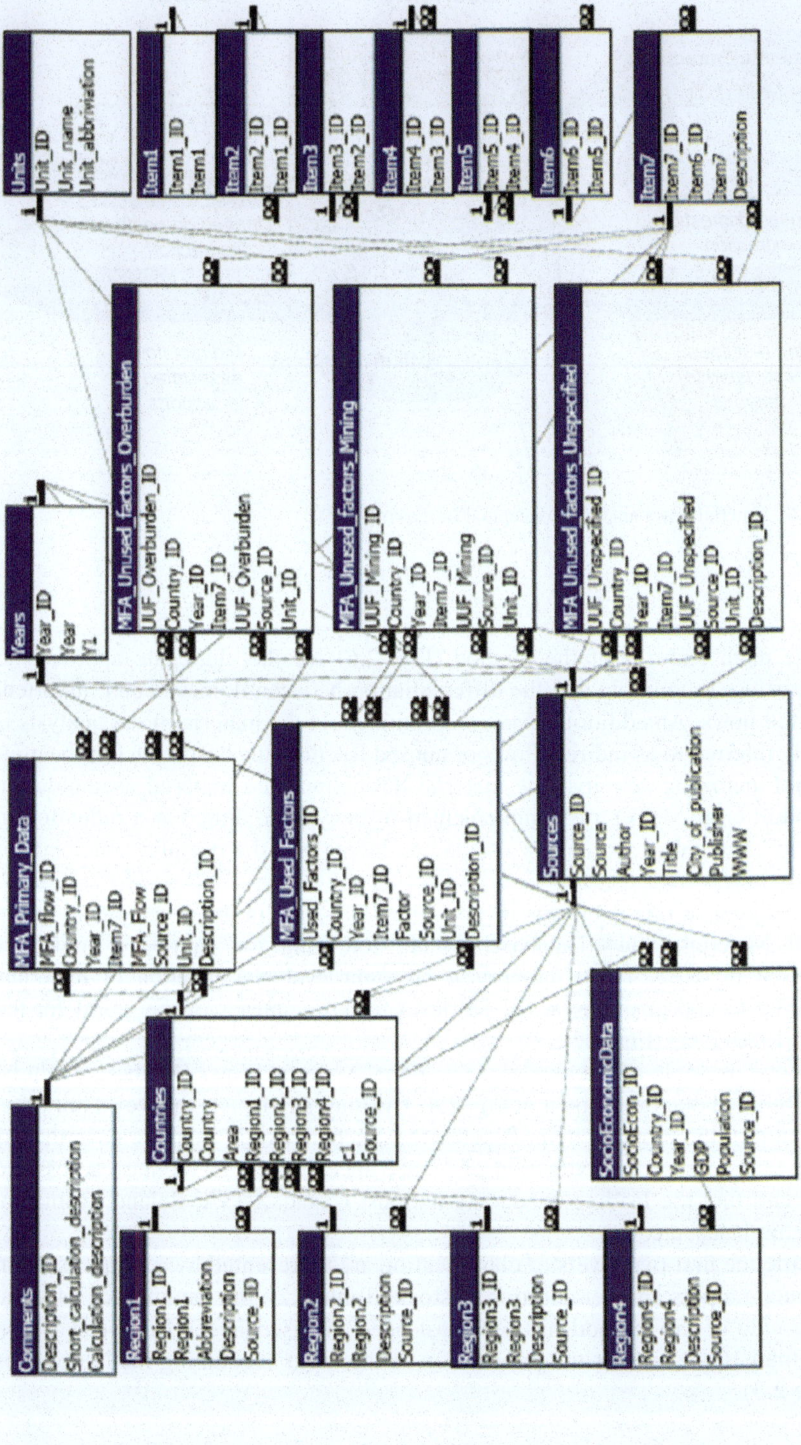

Fig. 2.7 Relationships of the global material flows database, (Shmelev and Giljum 2004)

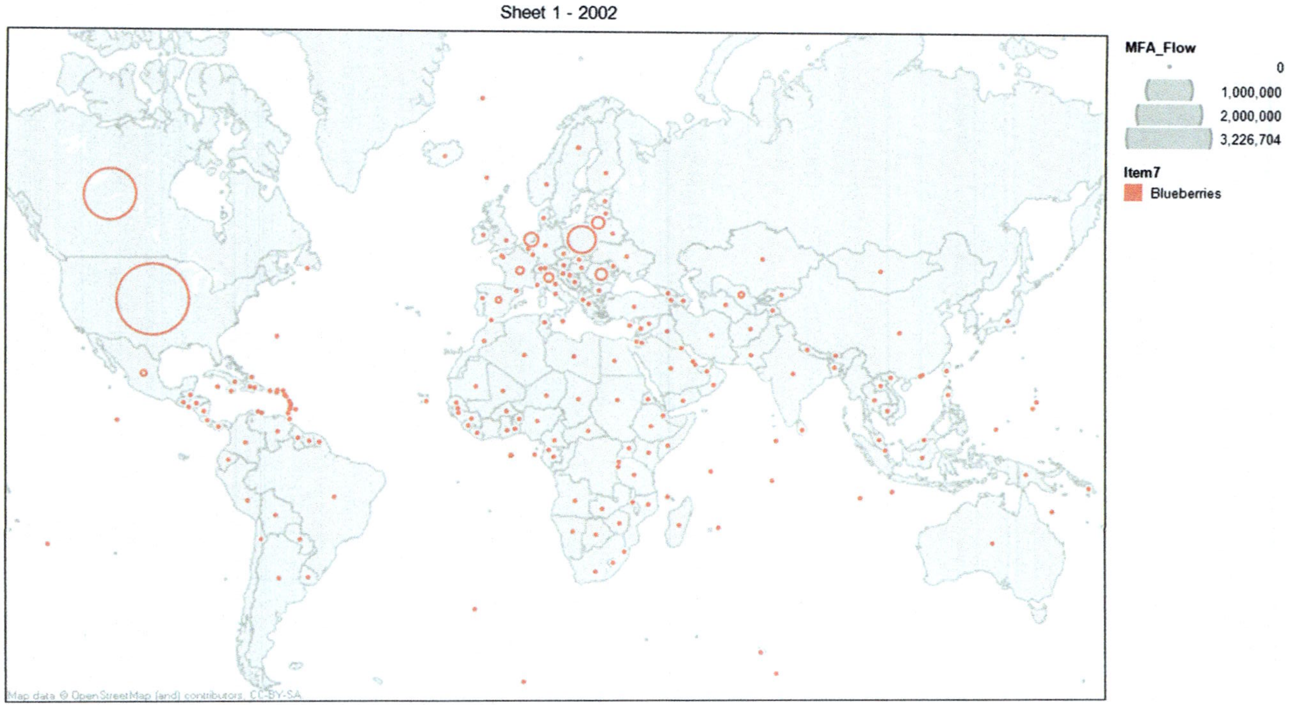

Map based on Longitude (generated) and Latitude (generated) on page 2002. Color shows details about Item7. Size shows sum of MFA_Flow. Details are shown for Country. The view is filtered on Item7, which keeps Blueberries.

Fig. 2.8 Global material flows database, domestic extraction, blueberries, 2002

Table 2.2 Global material flows analysis, top two levels, Shmelev, 2004

Material flows		
Item2_ID	Item1_ID	Item2
1	Input	Domestic extraction
2	Input	Imports
3	Output	Waste and emissions
4	Output	Dissipative use of products and dissipative losses
5	Output	Exports
6	Net addition to stock	Physical stocks

Table 2.3 Global material flows analysis, input and output, levels 2, 3 and 4, Eurostat and Shmelev, 2004–2010

Item 4_ID	Item 2	Item 3	Item 4
1	Domestic extraction	Fossil fuels	Hard coal
2	Domestic extraction	Fossil fuels	Lignite/brown coal
3	Domestic extraction	Fossil fuels	Crude oil
4	Domestic extraction	Fossil fuels	Natural gas
5	Domestic extraction	Fossil fuels	Natural gas liquids
6	Domestic extraction	Fossil fuels	Peat for energy use
7	Domestic extraction	Minerals	Metal ores
8	Domestic extraction	Minerals	Industrial minerals
9	Domestic extraction	Minerals	Construction minerals
10	Domestic extraction	Minerals	Industrial and construction minerals
11	Domestic extraction	Biomass	Biomass from agriculture
12	Domestic extraction	Biomass	Biomass from forestry
13	Domestic extraction	Biomass	Biomass from fishing
14	Domestic extraction	Biomass	Biomass from hunting
15	Domestic extraction	Biomass	Biomass from other activities
16	Domestic extraction	Fossil fuels	Other fossil fuels
17	Domestic extraction	Minerals	Other minerals
18	Domestic extraction	Biomass	Other biomass
19	Waste and emissions	Emissions to air	Carbon dioxide (CO_2)
20	Waste and emissions	Emissions to air	Methane (CH_4)
21	Waste and emissions	Emissions to air	Dinitrogen oxide (N_2O)
22	Waste and emissions	Emissions to air	Nitrous oxides (NO_x)
23	Waste and emissions	Emissions to air	Hydroflourcarbons (HFCs)
24	Waste and emissions	Emissions to air	Perflourocarbons (PFCs)
25	Waste and emissions	Emissions to air	Sulphur hexaflouride
26	Waste and emissions	Emissions to air	Carbon monoxide (CO)
27	Waste and emissions	Emissions to air	Non-methane volatile organic compounds (NMVOC)

(continued)

Table 2.3 (continued)

Item 4_ID	Item 2	Item 3	Item 4
28	Waste and emissions	Emissions to air	Sulphur dioxide (SO_2)
29	Waste and emissions	Emissions to air	Ammonia (NH_3)
30	Waste and emissions	Emissions to air	Heavy metals
31	Waste and emissions	Emissions to air	Persistent organic pollutants POPs
32	Waste and emissions	Emissions to air	Particles (e.g. PM_{10}, Dust)
33	Waste and emissions	Waste landfilled	Municipal waste
34	Waste and emissions	Waste landfilled	Industrial waste
35	Waste and emissions	Emissions to water	Nitrogen (N)
36	Waste and emissions	Emissions to water	Phosphorus (P)
37	Waste and emissions	Emissions to water	Heavy metals
38	Waste and emissions	Emissions to water	Other substances and (organic) materials
39	Waste and emissions	Emissions to water	Dumping of materials at sea
40	Dissipative use of products and dissipative losses	Dissipative use of products	Organic fertiliser (manure)
41	Dissipative use of products and dissipative losses	Dissipative use of products	Mineral fertiliser
42	Dissipative use of products and dissipative losses	Dissipative use of products	Sewage sludge
43	Dissipative use of products and dissipative losses	Dissipative use of products	Compost
44	Dissipative use of products and dissipative losses	Dissipative use of products	Pesticides
45	Dissipative use of products and dissipative losses	Dissipative use of products	Seeds
46	Dissipative use of products and dissipative losses	Dissipative use of products	Salt and other thawing materials spread on roads
47	Dissipative use of products and dissipative losses	Dissipative use of products	Solvents, laughing gas and other
48	Dissipative use of products and dissipative losses	Dissipative losses	Dissipative losses

Environmentally-Extended Input–Output Analysis

These ideas were clearly influenced by the work of Wassily Leontief in the field of input–output analysis in the USA economy (Leontief 1936, 1949, 1952), and especially by the environmentally extended applications of the input–output analysis to appear in (Leontief 1970, 1974, 1977a). Leontief built a conceptual link between the structure of the economy and the interdependent economic sectors and the environmental impacts of economic activity, namely air emissions.

Different countries started to develop input–output tables after the publication of the first balance of the national economy of the USSR and its subsequent criticism by Leontief. Tables for USA (1919, 1929, and 1947) followed. Later Norway (1948), the Netherlands (1948), Japan (1951) and the UK (1954) joined the process. With a little delay, Hungary (1957), Poland (1957), USSR (1959) and Brazil (1959) continued the trend. The resolution of the input–output tables varied significantly: if the first tables for the USA contained 44 and 41 sectors respectively, the Netherlands – 35 sectors; it was soon realised that increasing the amount of detail allows unprecedented capacity to understand and manage the complexity of intersectoral linkages. Subsequently tables for the USA included 400 sectors, Japan – 399 sectors; Estonia – 239 sectors; Lithuania – 239 sectors; Belorussia (500 sectors).

The first tables to appear in the USSR after WWII, including the tables for Estonia, Latvia and Lithuania (239 sectors, 1961), have been described in Jasny (1962) and Kossov (1964). The first Dutch input–output tables to appear have been reviewed by Rey and Tilanus (1963), the first international comparative analysis of the economies of the USA, Japan, Norway, Italy, Spain using input–output tables was offered by Simpson and Tsukui (1965).

Environmentally extended input–output applications started to develop in the 1970s following the original publication by Leontief and covered the following issues: energy and the environment (Carter 1974, 1976, Gay and Proops 1993, Herendeen and Tanaka 1976, Park 1982, Polenske and Lin 1993, Proops 1977, 1984); materials balance and materials flows (Duchin 2004, Giljum 2004, Hoekstra 2005, Suh 2009, Tukker et al. 2009); water (Anderson and Manning 1983, Dietzenbacher and Velázquez 2007, Lenzen 2009, Lenzen and Foran 2001, Wang and Wang 2009, Wang et al. 2005); waste (Duchin 1990, 1994, Kondo and Nakamura 2005, Leontief 1977b, Nakamura 1999, Nakamura and Kondo 2002, 2006) and environmental policy analysis (Gutmanis 1975). The UN global model project has significantly stimulated interest to the analysis of the environmental consequences of economic development and effects of technological innovation (Ayres and Shapanka 1976, Carter and Petri 1979, Leontief 1977c, Leontief and Duchin 1986, Petri 1977). Substantial projects focused on the application of input–output analysis to national economies for policy analysis have been started in various countries including the UK (Barker 1981, Barker et al. 1980, Stone 1984). Dynamic input–output analysis has become one of the most interesting subjects of economic research (Duchin and Szyld 1985, Raa 1986, Vogt et al. 1975). Environmentally extended input–output analysis of the changes in the world economy has been carried out by (Duchin 1986, Fontela 1989, Leontief and Duchin 1986, Schäfer and Stahmer 1989). Later, this framework was extended to include material flows (Duchin 2004), other pollutants (Duchin 1994, 1998) and different types of waste (Nakamura 1999). The most recent applications of extended input–output analysis today include an environmental key sector analysis by (Lenzen and Foran 2001), and econometric extended-input–output models of the UK and the European Union (Barker et al. 2007a, b).

The methods introduced in this chapter will be used in Chaps. 6, 7 and 11. As always, the interested reader is strongly advised to find and read the papers mentioned in this and subsequent chapters.

References

Anderson AW and Manning TW (1983) The Use of Input–Output Analysis in Evaluating Water Resource Development *Canadian Journal of Agricultural Economics* 31.1 15–26

Ayres RU (1978) *Resources, Environment and Economics. Applications of the Materials/Energy Balance Principle* New York, Wiley

Ayres RU (1995) Life Cycle Analysis: A Critique *Resources, Conservation and Recycling* 14.3–4 199–223

Ayres RU (2002) Industrial Metabolism and the Grand Nutrient Cycles in *Handbook of Environmental and Resource Economics* Cheltenham, Edward Elgar

Ayres RU and Ayres L (2002) *A Handbook of Industrial Ecology* Cheltenham, Edward Elgar

Ayres RU and Kneese AV (1969) Production, Consumption, and Externalities *The American Economic Review* 59.3 282–297

Ayres RU and Kneese AV (1971) Economic and Ecological Effects of a Stationary Economy *Annual Review of Ecology and Systematics* 2. 1–22

Ayres RU and Martinas K (1992) Experience and the Life Cycle: Some Analytic Implications *Technovation* 12.7 465–486

Ayres RU and Shapanka A (1976) Explicit Technological Substitution Forecasts in Long-range Input–Output Models *Technological Forecasting and Social Change* 9.1–2 113–138

Ayres RU and Simonis U (1994) *Industrial Metabolism: Restructuring for Sustainable Development* Tokyo, United Nations University Press

Ayres RU, D'Arge R and Kneese AV (1970) *Economics and the Environment: A Materials Balance Approach* Washington, Resources for the Future

Ayres RU, Ayres LW and MartÃnas K (1998) Exergy, Waste Accounting, and Life-Cycle Analysis *Energy* 23.5 355–363

Azapagic A and Clift R (1999) Life Cycle Assessment and Multiobjective Optimisation *Journal of Cleaner Production* 7.2 135–143

Barker T (1981) Projecting Economic Structure with a Large-Scale Econometric Model *Futures* 13.6 458–467

Barker T et al (1980) The Cambridge Multisectoral Dynamic Odel: An Instrument for National Economic Policy Analysis *Journal of Policy Modeling* 2.3 319–344

Barker T, Ekins P and Foxon T (2007) Macroeconomic Effects of Efficiency Policies for Energy-Intensive Industries: The Case of the UK Climate Change Agreements, 2000–2010 *Energy Economics* 29.4 760–778

Barker T, Junankar S et al (2007) Carbon Leakage from Unilateral Environmental Tax Reforms in Europe, 1995–2005 *Energy Policy* 35.12 6281–6292

Bengtsson M (2001) Weighting in Practice: Implications for the Use of Life-Cycle Assessment in Decision Making *Journal of Industrial Ecology* 4.4 47–60

Bouman M et al (2000) Material Flows and Economic Models: An Analytical Comparison of SFA, LCA and Partial Equilibrium Models *Ecological Economics* 32.2 195–216

Boustead I (1993) General Principles for Life Cycle Assessment Databases *Journal of Cleaner Production* 1.3–4 167–172

Carlson R et al (1998) Lci Data Modelling and a Database Design *The International Journal of Life Cycle Assessment* 3.2 106–113

Carter AP (1974) Energy, Environment, and Economic Growth *The Bell Journal of Economics and Management Science* 5.2 578–592

Carter AP (1976) *Energy and the Environment. A Structural Analysis* Waltham, Brandeis University Press

Carter AP and Petri PA (1979) Aspects of a New World Development Strategy II: Factors Affecting the Long-Term Prospects of Developing Nations *Journal of Policy Modeling* 1.3 359–381

De Udo Haes HA (1999) Weighting in Life-Cycle Assessment: Is There a Coherent Perspective? *Journal of Industrial Ecology* 3.4 3–7

Dietzenbacher E and Velázquez E (2007) Analysing Andalusian Virtual Water Trade in an Input–Output Framework *Regional Studies* 41.2 185–196

Duchin F (1986) Computers, Input–Output, and the Future *Journal of Economic Issues* 20.2 499–507
Duchin F (1990) The Conversion of Biological Materials and Wastes to Useful Products *Structural Change and Economic Dynamics* 1.2 243–261
Duchin F (1994) *Household Use and Disposal of Plastics an Input–Output Case Study for New York City* New York, New York University
Duchin F (1998) *Structural Economics: Measuring Change in Technology, Lifestyles and the Environment* Washington DC, Island Press
Duchin F (2004) *Input–Output Economics and Material Flows* Troy, Rensselaer Polytechnic Institute
Duchin F and Hertwich E (2003) *Industrial Ecology* Available at http://www.ecoeco.org/education_encyclopedia.php
Duchin F and Szyld DB (1985) A Dynamic Input–Output Model with Assured Positive Output(*) *Metroeconomica* 37.3 269–282
EEA (1999) *Material Flow-based Indicators in Environmental Reporting* Copenhagen, European Environment Agency
Fava JA (1997) LCA: Concept, Methodology, or Strategy? *Journal of Industrial Ecology* 1.2 8–10
Fischer-Kowalski M (1998) Society's Metabolism: The Intellectual History of Materials Flow Analysis, Part I, 1860–1970 *Journal of Industrial Ecology* 2.1 61–78
Fischer-Kowalski M and Hattler W (1998) Society's Metabolism: The Intellectual History of Materials Flow Analysis, Part II, 1970–1998 *Journal of Industrial Ecology* 2.4 107–136
Fontela E (1989) Industrial Structures and Economic Growth: An Input–Output Perspective *Economic Systems Research* 1.1 45–68
Gay PW and Proops JLR (1993) Carbon Dioxide Production by the UK economy: An Input–Output Assessment *Applied Energy* 44.2 113–130
Giljum S (2004) Trade, Materials Flows, and Economic Development in the South: The Example of Chile *Journal of Industrial Ecology* 8.1–2 241–261.
Gofman KG (2007) *The Work of Konstantin G. Gofman and Colleagues:An Early Example of Material Flow Analysis from the Soviet Union* Vienna, IFF
Graedel TE and Allenby BR (2002) *Industrial Ecology* Upper Saddle River, Prentice Hall
Guinée J and Heijungs R (1993) A Proposal for the Classification of Toxic Substances Within the Framework of Life Cycle Assessment of Products *Chemosphere* 26.10 1925–1944
Guinée JB et al (1993) Quantitative Life Cycle Assessment of Products: 2. Classification, Valuation and Improvement Analysis *Journal of Cleaner Production* 1.2 81–91
Gutmanis I (1975) Input–Output Models in Economic and Environmental Policy Analyses *Proceedings of the IEEE* 63.3 431–437
Haes HAU et al (2004) Three Strategies to Overcome the Limitations of Life-Cycle Assessment *Journal of Industrial Ecology* 8.3 19–32
Hanssen OJ and Asbjørnsen OA (1996) Statistical Properties of Emission Data in Life Cycle Assessments *Journal of Cleaner Production* 4.3–4 149–157
Heijungs R and Guinée J (1993) Software as a Bridge Between Theory and Practice in Life Cycle Assessment *Journal of Cleaner Production* 1.3–4 185–189
Heijungs R and Suh S (2002) *The Computational Structure of Life Cycle Assessment* Dordrecht/Boston/London, Kluwer Academic Publishers
Heijungs R and Suh S (2006) Reformulation of Matrix-Based LCI: From Product Balance to Process Balance *Journal of Cleaner Production* 14.1 47–51
Heijungs R and United Nations Environment Programme (1996) *Life Cycle Assessment: What it is and How to Do it* (1st edition) Paris, United Nations Environment Programme
Herendeen R and Tanaka J (1976) Energy cost of living *Energy* 1.2 165–178
Hoekstra R (2005) *Economic Growth, Material Flows and the Environment. New Applications of Structural Decomposition Analysis and Physical Input–Output Tables* Cheltenham, Edward Elgar
International Organization for Standardization (2007) *ISO 14000, Environmental Management* (1st edition) Geneva, ISO
Jasny N (1962) The Russian Economic "Balance" and Input–Output Analysis: A Historical Comment *Soviet Studies* 14.1 75–80

Kondo Y and Nakamura S (2005) Waste Input–Output Linear Programming Model with Its Application to Eco-Efficiency Analysis *Economic Systems Research* 17.4 393–408

Kossov VV (1964) Regional Input–Output Analysis in the U.S.S.R. *Papers in Regional Science* 14.1 175–181

Lenzen M (2009) Understanding Virtual Water Flows: A Multiregion Input–Output Case Study of Victoria *Water Resources Research* 45.9 W09416

Lenzen M and Foran B (2001) An Input–Output Analysis of Australian Water Usage *Water Policy* 3 321–340

Leontief W (1936) Quantitative Input and Output Relations in the Economic Systems of the United States *The Review of Economics and Statistics* 18.3 105–125

Leontief W (1949) Recent Developments in the Study of Interindustrial Relationships *The American Economic Review* 39.3 211–225

Leontief W (1952) Some Basic Problems of Structural Analysis *The Review of Economics and Statistics* 34.1 1–9

Leontief W (1970) Environmental Repercussions and the Economic Structure: An Input–Output Approach *The Review of Economics and Statistics* 52.3 262–271

Leontief W (1974) Structure of the World Economy. Outline of a Simple Input–Output Formulation *The Swedish Journal of Economics* 76.4 387–401

Leontief W (1977a) Natural Resources, Environmental Disruption, and Growth Prospects of the Developed and Less Developed Countries *Bulletin of the American Academy of Arts and Sciences* 30.8 20–30

Leontief W (1977b) Natural Resources, Environmental Disruption, and the Future World Economy *Journal of International Affairs* 31.2 267

Leontief W (1977c) *The Future of the World Economy: A United Nations Study* New York, Oxford University Press

Leontief W and Duchin F (1986) *The Future Impact of Automation on Workers* New York, Oxford University Press

Lovelock JE (1972) Gaia as Seen Through the Atmosphere *Atmospheric Environment (1967)* 6.8 579–580

Margulis L and Lovelock JE (1978) The Biota as Ancient and Modern Modulator of the Earth's Atmosphere *Pure and Applied Geophysics* 116.2 239–243

Moiseev NN (1978) Cybernetic Description of Ecological-Economic Systems *Cybernetics* 13.6 930–944

Moiseev et al (1983) Global Models, the Biospheric Approach (Theory of the Noosphere) [accessed 18 Mar 2010] Available at http://www.iiasa.ac.at/Admin/PUB/Documents/CP-83-033.pdf

Moiseev NN, Alexandrov VV and Tarko AM (1985) *Chelovek i Biosfera. Opyt Sistemnogo Analiza i Eksperimenty s Modelyami* Moscow, Nauka [accessed 18 Mar 2010] Available at http://www.ras.ru/win/db/show_ref.asp?P=.id-25414.ln-en

Nakamura S (1999) An Interindustry Approach to Analyzing Economic and Environmental Effects of the Recycling of Waste *Ecological Economics* 28.1 133–145

Nakamura S and Kondo Y (2002) Recycling, Landfill Consumption, and CO_2 Emission: Analysis by Waste Input–Output Model *Journal of Material Cycles and Waste Management* 4.1 2–11

Nakamura S and Kondo Y (2006) A Waste Input–Output Life-Cycle Cost Analysis of the Recycling of End-of-Life Electrical Home Appliances *Ecological Economics* 57.3 494–506

Park S-H (1982) An Input–Output Framework for Analysing Energy Consumption *Energy Economics* 4.2 105–110

Petri PA (1977) An Introduction to the Structure and Application of the United Nations World Model *Applied Mathematical Modelling* 1.5 261–267

Polenske KR and Lin X (1993) Conserving Energy to Reduce Carbon Dioxide Emissions in China *Structural Change and Economic Dynamics* 4.2 249–265

Proops JLR (1977) Input–Output Analysis and Energy Intensities: A Comparison of Some Methodologies *Applied Mathematical Modelling* 1.4 181–186

Proops JLR (1984) Modelling the Energy–Output Ratio *Energy Economics* 6.1 47–51

Raa TT (1986) Applied Dynamic Input–Output with Distributed Activities *European Economic Review* 30.4 805–831

Rey G and Tilanus CB (1963) Input–Output Forecasts for the Netherlands, 1949–1958 *Econometrica* 31.3 454–463

Schäfer D and Stahmer C (1989) Input–Output Model for the Analysis of Environmental Protection Activities *Economic Systems Research* 1.2 203–228

Simpson D and Tsukui J (1965) The Fundamental Structure of Input–Output Tables, An International Comparison *The Review of Economics and Statistics* 47.4 434–446

Stone R (1984) Model Design and Simulation *Economic Modelling* 1.1 3–23

Suh S (ed) (2009) *Handbook of Input–Output Economics in Industrial Ecology* New York, Springer

Tukker A et al (1997) Combining SFA and LCA: The Swedish PVC Analysis *Journal of Industrial Ecology* 1.4 93–116

Tukker A et al (2009) Towards a Global Multi-Regional Environmentally Extended Input–Output Database *Ecological Economics* 68.7 1928–1937

United Nations (1947) *Measurement of National Income and the Construction of Social Accounts* Geneva, United Nations

United Nations (1953) *System of National Accounts and the Supporting Tables*

United Nations (1968) *A System of National Accounts* New York, United Nations

United Nations (1993a) *Integrated Environmental and Economic Accounting*

United Nations (1993b) System of National Accounts 1993

United Nations (2003) *Handbook of National Accounting: Integrated Environmental and Economic Accounting 2003*

United Nations (2009) *System of National Accounts 2008* Available at http://unstats.un.org/unsd/nationalaccount/SNA2008.pdf

Vernadsky VI (1924) *La Geochimie* Paris, Librairie Félix Alcan

Vernadsky VI (1926) *Biosphera* Nauchnoe Chimico-Technicheskoe Izdatelstvo Moscow, Leningrad

Vernadsky VI (1929) *La Biosphere* Paris, Librairie Félix Alcan

Vernadsky VI (1940) *Biogeochimicheskie ocherki* Moscow, Leningrad

Vogt WG, Mickle MH and Aldermeshian H (1975) A Dynamic Leontief Model for a Productive System *Proceedings of the IEEE* 63.3 438–443

Wang H and Wang Y (2009) An Input–Output Analysis of Virtual Water Uses of the Three Economic Sectors in Beijing *Water International* 34.4 451–467

Wang L MacLean HL and Adams BJ (2005) Water Resources Management in Beijing Using Economic Input-Output Modeling *Canadian Journal of Civil Engineering* 32 753–764

Weizsäcker E and Club of Rome (1998) *Factor Four: Doubling Wealth – Halving Resource Use: The New Report to the Club of Rome* London, Earthscan Publications Ltd.

Chapter 3
The Big Picture Vision and the Environment: An International Perspective

Abstract Because environmental problems usually manifest themselves on a global scale, global modelling tools are needed to study them and to design possible solutions. Starting from a concept of global biogeochemical cycles, proposed by Vernadsky, through a series of insights provided by the modellers, such as Moiseev, Naess, Lovelock, the chapter approaches the criteria for judging global models. Several major contributions to the global modelling literature are reviewed and models are compared on a range of parameters, such as the most frequently used modelling technique, the presence of the time dimension, the treatment of population change, energy, agriculture, prices, trade and environmental pollution. Drawn from the experience of D. Meadows and accompanied by the charts, depicting various dimensions of EU development on the regional scale, this chapter is designed to introduce the subject of global modelling to readers, who will find a plethora of additional literature focused on this subject following the references provided.

Keywords Global modelling • Pollution • Systems dynamics • Optimization • Input–output

Vladimir Vernadsky and "Geochemistry"

The study of environmental-economic interactions cannot avoid difficult issues of choice and the wider question of worldview, which either helps to live in harmony with the natural world or determines its destruction. One can search for the roots of environmental ethics in Chinese Dao thinking; however, if we concentrate on the twentieth century, there are several key figures we will need to explore. One of the first scientists to realise the role of humans as a geological force on our planet was the Russian geochemist Vladimir Vernadsky (1863–1945). Vernadsky became interested in global biogeochemical cycles and essentially started a new discipline of geochemistry in his book "Geochemistry" (Vernadsky 1924), which later evolved into Earth Systems Science. In his second major book, "The Biosphere" which,

S.E. Shmelev, *Ecological Economics: Sustainability in Practice*,
DOI 10.1007/978-94-007-1972-9_3,

after being published in Russian was issued in French (Vernadsky 1929) Vernadsky explores the idea of the biosphere, first introduced by the Austrian geologist Eduard Suess. In his later writings (Vernadsky 1936) he explored the idea of the noosphere, the evolution of humanity's scientific thought, which becomes so powerful with the modern discoveries of physics that it starts playing a leading role in the evolution of the Earth system (Lapo 2001). The first full English translation of Vernadsky's main work appeared in 1986 (Vernadsky 1986). Vernadsky was a member of the St Petersburg school, to which the soil scientist Vassily Dokuchaev (1846–1903) (Dokuchaev 1879) and the creator of the periodic table of chemical elements Dmitry Mendeleev (1834–1907) (Mendeleev 1869, 1871) also belonged.

Aldo Leopold and "Land Ethic"

The landmark figure in American environmental thought was Aldo Leopold, who in his book "Sand County Almanac" (Leopold 1949) expressed a great passion for the preservation of landscape and the change of emphasis on land as a commodity, which we owe to the community to which we belong. Aldo Leopold realised that resource management could cause damage to the wider environment and his Land Ethic demanded environmental management as a new form of more inclusive interaction with the land. Leopold viewed the ecosystem as a complex system, a pyramid of species, exchanging energy; and he thought that the goal of management should be to protect the integrity of the system (Norton 1990).

Rachel Carson and "Silent Spring"

The next major step in the development of ecological consciousness (Shmeleva 2006) came in the form of a book "Silent Spring" (Carson 1962) by Rachel Carson, which focused on the use of chemicals in agriculture, especially DDT. It has been shown that, through agricultural use, mixing with water, the chemicals reach the sea and through the food chain appear as far from their origin as in the bodies of penguins. The book attracted much media attention and proved highly influential for the environmental movement.

Donella and Dennis Meadows and "Limits to Growth"

Donella Meadows and Dennis Meadows at MIT explored the idea of the carrying capacity of the earth and bridged the gap between mathematical modelling and ethics by offering a conceptual foundation for large scale systems analysis (Meadows and Club of Rome 1972, Randers and Donella 1972). These works were a major step towards initiating a serious scientific discussion on the carrying capacity of the earth, the limits to unconstrained economic growth, causing depletion of resource stocks, and environmental pollution. The sequels, (Meadows 1992, 2005) provide an update on the modelling and conceptualisation issues related to global sustainability analysis.

In her book "Groping in the Dark" (Meadows 1982) Donella Meadows developed a useful set of questions referring to global modelling which it seems would be appropriately cited here. This questionnaire was distributed among the leading global modellers of the time and provides an invaluable help in creating all kinds of conceptual models of the world: mathematical, verbal and structural, with a systems approach in mind.

Box 3.1 Global Modelling Questionnaire

1. **The purposes and goals of global modelling**

What are the main problems (what is the single most important problem) a global model should try to analyse? To what extent can global modelling serve this purpose?

What are the specific features of goal-setting in a global modelling effort?

How should normative aspects interrelate with descriptive aspects? What should be the predictive value of a global model? Under what conditions? If no absolute prediction is possible, how should the model be put to use?

What kind of 'global goal function' could be conceived or what other possibilities of representation of goal-seeking behaviour do you see in global modelling?

How would you proceed if you had (a) limited resources or (b) (practically) unlimited resources to spend on global modelling work?

What services might (will) global modelling be able to render in the future?

What services not?

2. **Methodology**

Given the purposes spelled out under part 1, what are the consequences with respect to methodology? For model structure? For mode of model use? How far do certain methodologies reflect/determine certain world views?

In particular:

- How should aspects be handled that one knows to be important but about which one lacks data or knowledge of relations (for example environment or 'the human factor')?
- Many modellers have rather uncritically used cross-sectional, static data to estimate what are essentially longitudinal, dynamic relationships. Do you perceive this as a major problem? Are there strategies for getting around it or must we live with the constraint of insufficient time-series data?

(continued)

Box 3.1 (continued)

- How can non-material needs be represented?
- How can physical constraints be best treated (limiting values, for example the proportion of carbon dioxide in the atmosphere, the rate of heat dissipation, etc.)?
- How should the price system be taken care of?

3. **Actors, policy variables**

Who are (a) the actors in the model whose behaviour is endogenously modelled, (b) the actors steering the model, and how are they represented?

How does the model handle policy variables? Does the model assume that policies remain constant unless specific policy intervention is made during the course of the run? Or does the model automatically change policy variables during the course of a run, thus giving the appearance that problems have vanished without any specific action on the part of national leaders? What policy options can the model test?

How do the model policy options relate to the policy options and choices being made by leaders today? How many of the policy variables correspond to policy alternatives that political decision makers can actually select, given their power, and how many are of a more abstract, synthetic nature?

How should political constraints of an institutional kind be tackled methodologically?

What do the structural equations depict in terms of goals of actors?

How do the actors interact?

4. **Structural aspects**

Regarding past modelling efforts, could one have arrived at the same results (i) with a smaller model or (ii) with a different model (approach)?

Have you developed a formalism by which to decide what to leave out of a model?

Have you made any effort, after the model was constructed, to compress and simplify its essence in a form that would allow for understanding rather than simply complex explanation? What insights does the model yield that would not have been available through other means of analysis?

How is consistency among regions guaranteed?

How does the model represent the fact that actors adapt to changes?

How should changes in structural relationships be considered? How should technological progress/technological change be represented? What other types of structural relationships are important in this context? Could 'structure sensitivity analysis' help to answer these questions?

Should a model be set up as a one-time venture or should it deliberately be designed in a modular fashion, flexible enough to adapt to new insights and/or new policy questions? How can such flexibility be ensured?

(continued)

Box 3.1 (continued)

5. **Testing the model**

Which aspects of your model do you have much confidence in and which aspects do you have less confidence in? Which aspects of your model require the most subjective judgement? Which aspects are the most concrete? What kind of errors are to be expected (size and frequency)? What assumptions in your model are you least certain about?

What provision have you made to take the uncertainty of the parameters into consideration and the uncertainty in the data and in the structural equations? Has your model produced sensible results when subjected to noise and larger disturbances?

How can one be assured that the model will correspond to reality (validation)?

How can one be assured that the model did correspond to reality (calibration)?

What is the role and purpose of sensitivity tests?

6. **Internal organisation**

How should global modelling work be organised?

How can consistency between subgroups be secured?

How should global modelling groups cooperate in the future?

What rules and procedures would you lay down to ensure that documentation of the model is kept continuously up to date?

What are the best ways of interweaving with the scientific community at large?

7. **Relations between modeller and user**

Should an explicit client be identified at an early stage?

To what extent did clients participate in the work? What formats to communicate with the clients should be developed?

What can be done to help the client to avoid misunderstandings and to understand the strengths, weaknesses and limitations of the model?

Which clients need world modelling most? Which are most interested in world modelling?

Do you have recommendations on how to reduce 'overselling'?

Global Modelling

Since the time of publication of the "Limits to Growth" several world laboratories have undertaken major projects on the modelling of the global system (Table 3.1). Among them the works on the Japanese FUGI model by Onishi (Onishi 1977, 2001,

2005, 2009); a South American normative Bariloche model (Gallopin 2001, Scolnik 1979); the UK Government SARUM model (Roberts 1977), the UN world model developed by Leontief and his team (Leontief 1974, 1977a, b); the German GINFORCE model: (Lutz 2010, Lutz and Meyer 2009), and the Cambridge E3MG model (Koehler et al. 2006). Table 3.1 presents the outline of dominant approaches used in models of the world economy.

It should be mentioned that the scope of the models, their goals, the assumptions embedded in them, and the variables that are taken into account differ profoundly. Table 3.2 compares the models outlined above from the point of view of the aspects of reality reflected in them.

James Lovelock and "Gaia Theory"

James Lovelock, addressing the issue of criteria for existence of life on other planets (Lovelock 1972, 1990, Lovelock and Whitfield 1982, Margulis and Lovelock 1978) formulated a hypothesis of the self-regulating nature of the biosphere, especially in the context of maintaining a certain concentration of greenhouse gases. These views resulted in a book "Gaia, a new look at life on earth" (Lovelock 1979), which has since seen many editions. The formulation of the Gaia hypothesis has met with a lot of criticism and opposition, although the idea is certainly very close to those of Vernadsky, Moiseev and the biogeochemistry school. The common themes were explored in the Biosphere and Noosphere Reader published by Routledge (Samson and Pitt 1999). More recent works on the subject include (Kleidon 2010, Worden 2010).

Nikita Moiseev and "Ecological-Economic Modelling"

Nikita Moiseev, working on the range of ecological-economic modelling problems (Moiseev 1978, 1982, 1994; Moiseev et al. 1983) drew on the theory of the noosphere by V. Vernadsky. Writing in Russian, Moiseev explored the issues of interaction between the economy and the environment (Moiseev et al. 1985), global modelling (Moiseev 1988), and changing climatic conditions as a result of a hypothetical nuclear conflict (Moiseev 1987b). Moiseev arrived at the conclusion that the Earth would become uninhabitable if nuclear weapons were ever employed on a large scale. This led to the development of Moiseev's concept of the ecological imperative (Moiseev 1987a) and influenced much environmental thinking in the Russian-speaking world. His "Reflection on the Noosphere. Humanism of Our Time" written in English was published in the Biosphere and Noosphere Reader (Samson and Pitt 1999).

Table 3.1 Modelling paradigms in world models: X – primary influence; (X) – secondary (Source: Meadows (1982) influence)

The paradigm	Forrester & Meadows	Pestel & Mesarovic	Bariloche	MOIRA	SARUM	FUGI	UN Leontief	E3MG	GINFORCE
Systems dynamics	X	(X)			(X)	(X)			
Economics/econometrics			(X)	X	X	(X)		X	X
Optimization		(X)	X	(X)					
Input–output analysis						X	X	X	X
Eclectic	X	(X)						(X)	(X)

Table 3.2 Comparison of the global models

Variable class or sector	Forrester & Meadows	Pestel & Mesarovic	Bariloche	MOIRA	SARUM	FUGI	UN Leontief	E3MG	GINFORCE
Population	X	X	X	X	X	X	X		
Energy and mineral resources	X	X			X	X	X	X	X
Agricultural sector:									
Consumption			X	X	X	X	X	X	X
Production	X	X	X	X	X	X	X	X	X
Prices/market mechanism		X		X	X	X	X	X	X
Trade		X	X	X	X	X	X	X	X
Environmental pollution	X						X	X	X

Arne Naess and "Deep Ecology"

Arne Naess, interestingly, working in Norway on similar issues of mathematical modelling of conflicts arrived at similar conclusions and became an active proponent of a concept of deep ecology (Naess 1973, 1993, 2008).

Two international journals, Environmental Ethics (started in 1979) and Environmental Values (started in 1992) provided a forum for the exchange of ideas on the ethical dimensions of environmental change, interaction between humans and the environment, the notion of sustainability and decision making.

The discussion on the notion and the ethical dimension of the sustainable development concept accelerated after the World Summit on Sustainable Development in Rio de Janeiro in 1992 (Beckerman 1994, Daly 1990a, b, Pezzey 1992). David Pearce suggested a new programme of Green Economics (Pearce 1992) and the discussion on sustainability continued (Common and Perrings 1992, Goodland and Daly 1996).

Incommensurability of Values

Incommensurability of values has been another hotly debated ethical issue over recent decades. It is generally accepted that "alternatives are incommensurable when they cannot be precisely measured along some common cardinal scale of units of value" (Aldred 2006). The idea was first introduced in (Griffin 1977) and later developed, debated and defended in (Adler 1998, Aldred 1994, 2002, 2006; Arrow 1997, Attfield 1998, Holland 1994, Martinez-Alier et al. 1998, O'Neill et al. 2007, Vatn 2000). Using incommensurability of values as the guiding principle, the recent attempts to put a value on climate change effects by Stern Spash (2007) and biodiversity (Spash 2008) have been criticised.

To summarise them: the ideas of holism and interdisciplinarity, a systems approach, developments in physics: the concept of the arrow of time and the second law of thermodynamics, the idea of the ecological imperative, of deep ecology and environmental stewardship, and of incommensurability of values, brought to economics from various neighbouring disciplines including philosophy, mathematics and physics, addressed logical inconsistencies of the theory and created a basis for the application of more sophisticated tools, including multicriteria decision aid (the subject of next chapter), environmentally extended input–output analysis (the subject of Chap. 6) and many other approaches.

Distribution of global GDP per capita (Fig. 3.1) gives us a global picture of inequalities in the levels of economic development existing in the world. There is a significant degree of correlation between the level of GDP per capita and CO_2 emissions per capita, which is largely related to the lifestyle choices people make (individual decisions) and the technological options they have (group decisions).

Fig. 3.1 Distribution of global GDP

Spatial Element of Economy-Environment Interactions

In order to understand fully the causes of the unsustainability of our development mechanisms and trends and the necessary steps to transfer to the sustainable development trajectory we need to understand another important element of economy-environment interactions: the spatial element. By having big picture vision we create understanding of the global system and the patterns present in it: where major resources are extracted, where they are processed, where the final products tend to be consumed; which technologies of environmental protection are better for which geographical and climatic conditions, which lifestyle choices are available to people in different parts of the world, how sustainability might be embedded in the fabric of urban life in various corners of the world, how to make a global transportation network more sustainable, how to resolve the conflict between the preservation of ecosystems and biodiversity and poverty as well as the need for development, how to develop more environmentally sustainable modes of agricultural production and how to improve the social aspects of working in cities.

Analysis of land cover and land cover change (Fig. 3.2) could shed light on the economy-environment interaction in biomass-intensive sectors, such as agriculture and forestry, the dominant land-users in the national economy, as well as on urbanisation patterns and location choices for enterprises. Taking into account the location of valuable ecosystems should be a priority when designing new transport routes,

Fig. 3.2 European land cover

carrying out construction projects, managing waste. Also, studying the spatial disaggregation at levels lower than the national helps us to understand the link between different aspects of the economy and causes of successful development (in Europe such regions are called NUTS, "Nomenclature of Territorial Units for

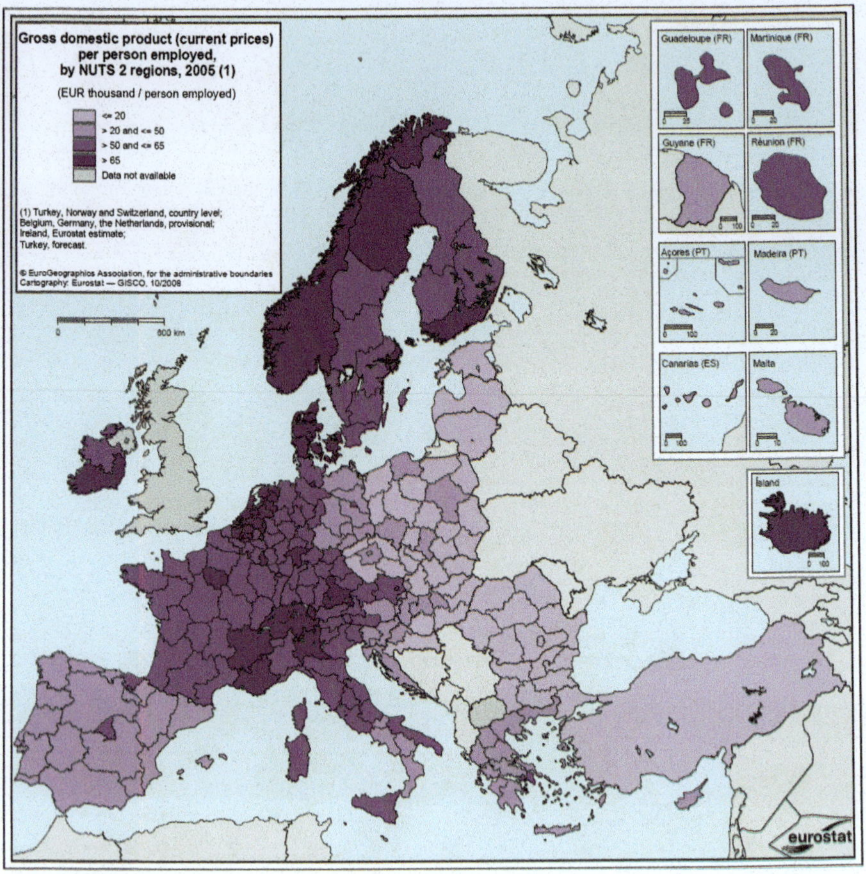

Source: Eurostat (reg_e2gdp)

Fig. 3.3 EU GDP per person employed, NUTS2 regions, 2005

Statistics"). The quality of the regional statistics in the EU is constantly improving. Apart from the Gross Regional Product per capita (Fig. 3.3) one can see regional data for unemployment (Fig. 3.4), R&D expenditure (Fig. 3.5), Number of students in Higher Education (Fig. 3.6), and dominant sectoral structure (Fig. 3.7). All this data completes the picture obtained through information at the national level and might provide valuable insights into possible mechanisms of sustainability interventions and policy improvements.

Figure 3.8 will provide the basis for the chapter on Sustainable Cities in this book by highlighting the location of predominantly urban and rural territories in Europe (urban are coloured in red). Such a spatial perspective is extremely useful in explaining the forces of economic development as well as the patterns of material and energy flows within Europe. Historically cities have been the centres of accumulation of knowledge, crafts and technology. The concentration of talented people

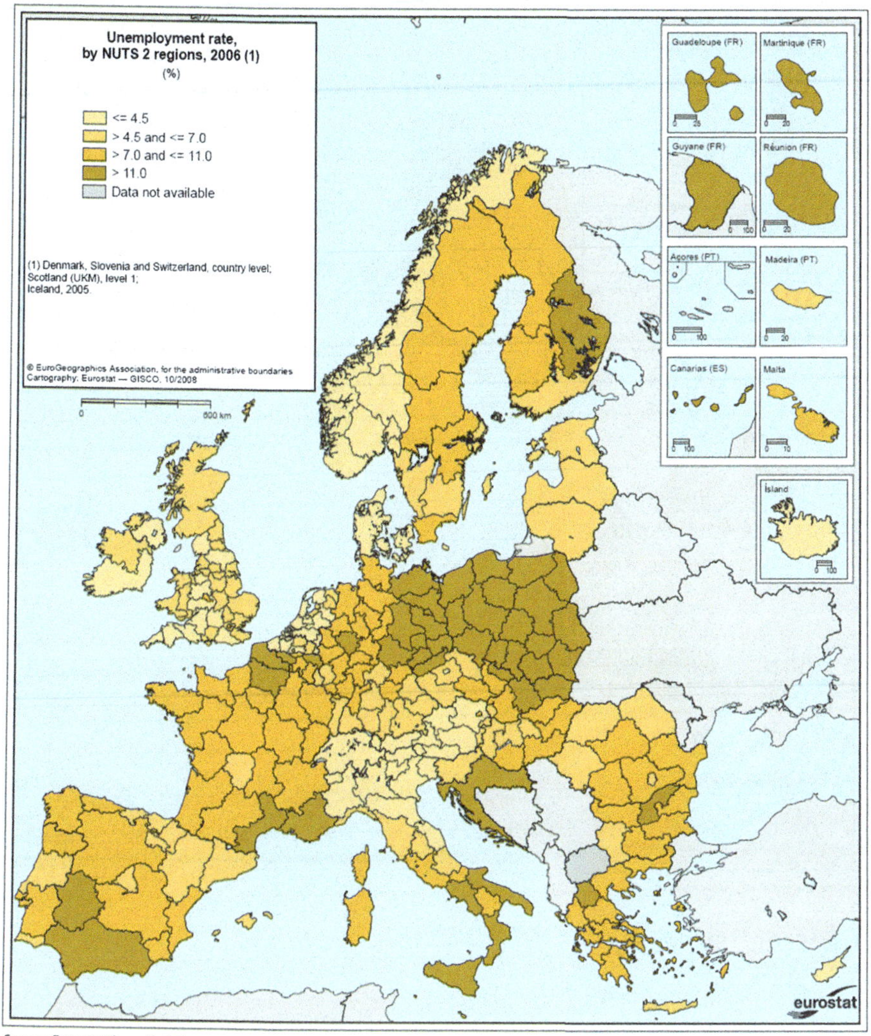

Source: Eurostat (reg lfu3rt)

Fig. 3.4 Unemployment rates, EU NUTS2 regions, 2006

led to further improvements in the quality of manufactured products and led to the inflow of additional craftsmen and entrepreneurs. Several cities positioned close to each other amplified the effect. This led to the formation of the so-called "core" regions of Europe: London and the South of England, Paris and the North of France, Belgium, the Netherlands, part of Western Germany. This region started to accumulate more wealth, especially after the development of maritime trade routes after the Age of Discoveries. Even today, it provides considerable employment, generates a larger share of GDP than its neighbours and acts as a driver for

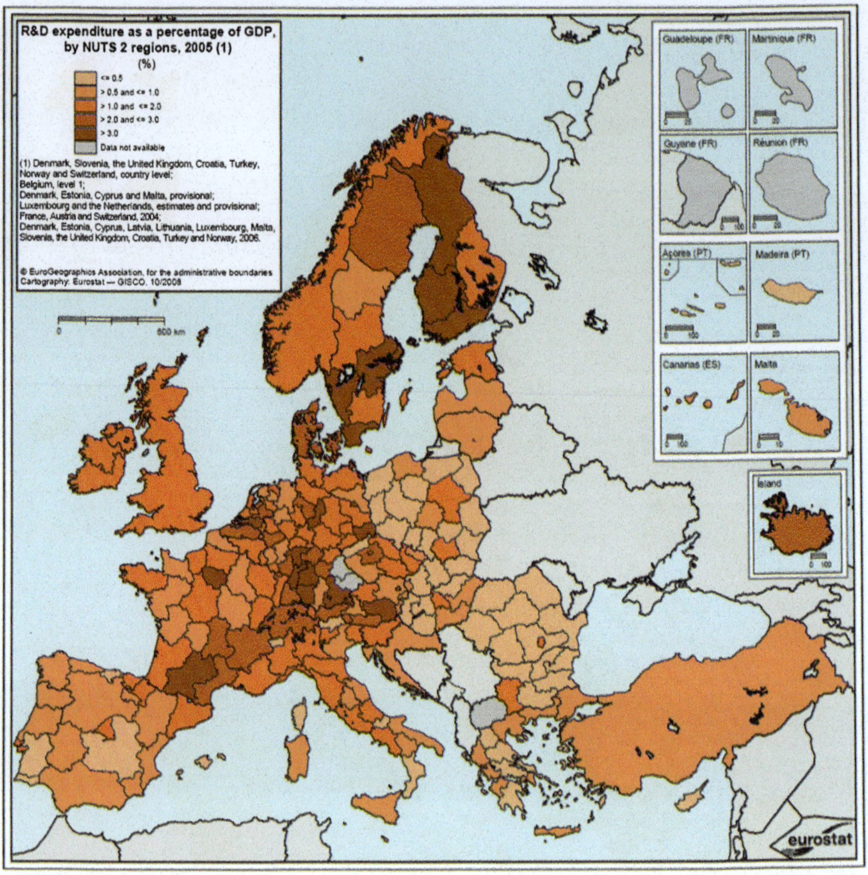

Source: Eurostat (rd_e_gerdreg)

Fig. 3.5 R&D expenditure as a % of GDP, 2005

economic development for the whole of Europe. On the other hand, it is here that substantial amounts of energy are used, much water is consumed by residents, and great deal of food and resources are transported here from the surrounding regions as well as from abroad. By observing and studying these flows we can better understand the processes of economic development and the impacts they make on the environment, for such a representation brings us closer to reality than do aggregated macroeconomic models.

When exploring large international datasets, like the one devoted to the pattern of the consequences of the financial crisis in the EU depicted in Figs. 3.9 and 3.11, a strategy of using cob-web diagrams proves to be extremely useful as it allows us

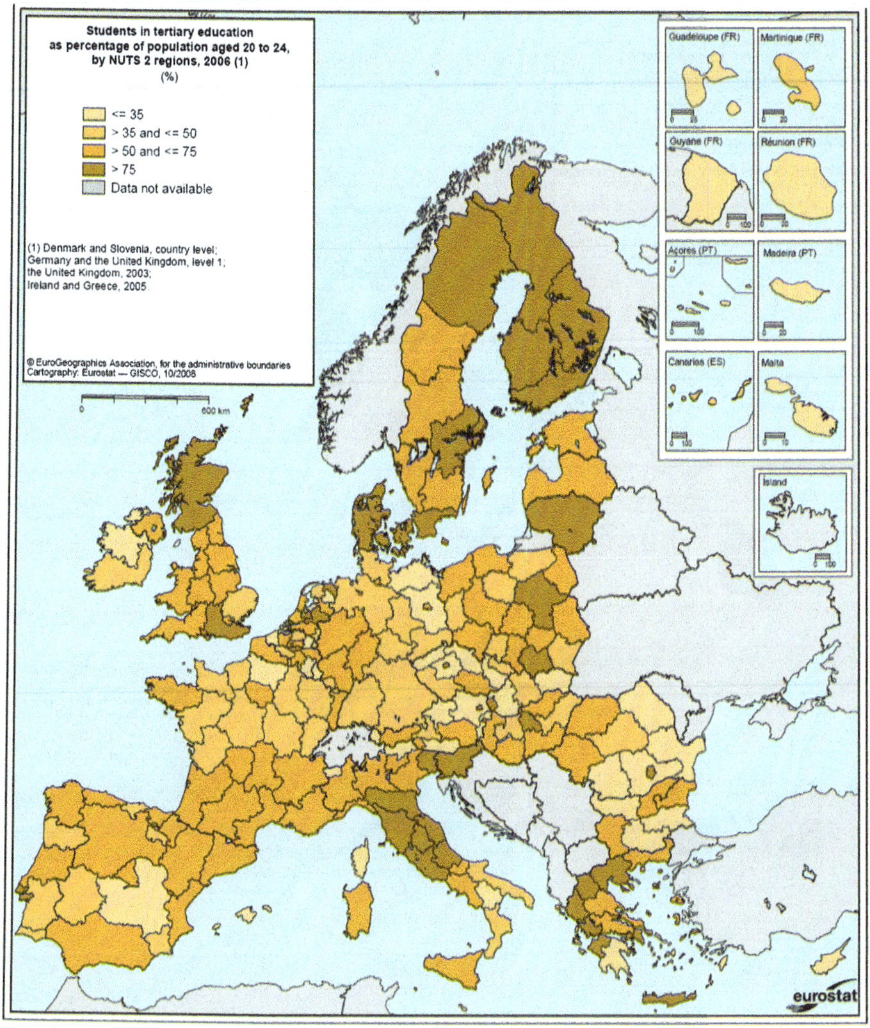

Source: Eurostat (educ_renrlrg1 and reg_d2jan)

Fig. 3.6 Students in tertiary education, NUTS2 regions, 2006

to demonstrate the values of a particular variable in many locations simultaneously, even adding the time dimension if the data allows it.

As can be seen in Fig. 3.9, unemployment increased in the majority of EU member states as a result of a financial crisis with the cases of Spain (18%), Latvia (almost 18%), Estonia (14%), Hungary (10%), Portugal (10%), and Slovakia (12%) being the most severe.

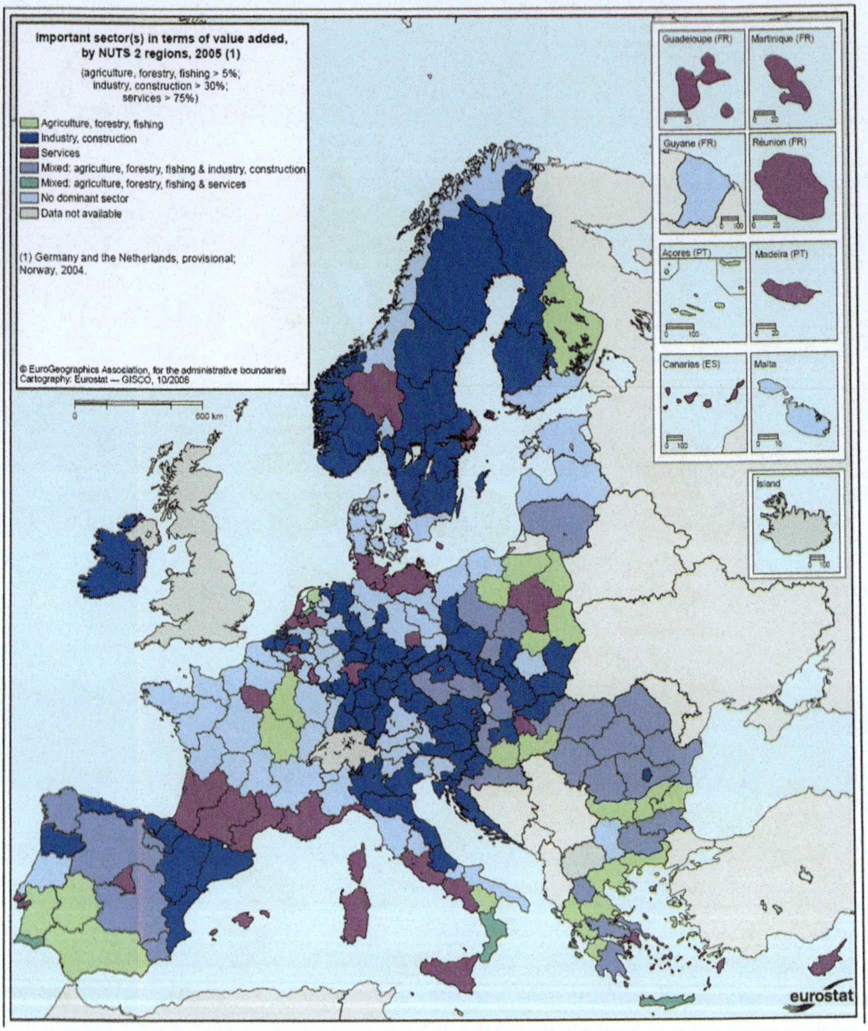

Source: Eurostat (req_e3vabp95)

Fig. 3.7 Important EU sectors in terms of value added, NUTS2, 2005

From Fig. 3.10 one can infer that the budget deficit was most severe in Greece, the UK, Ireland, Spain, Portugal, Latvia and Lithuania. Figure 3.11 shows that tackling inflation was chosen as a priority and many governments managed to bring it down. Most considerable reductions could be seen in Bulgaria, Estonia, Latvia and Lithuania.

This chapter highlights the importance of big picture vision; a discussion of systems thinking and philosophical approaches to understanding global sustainability

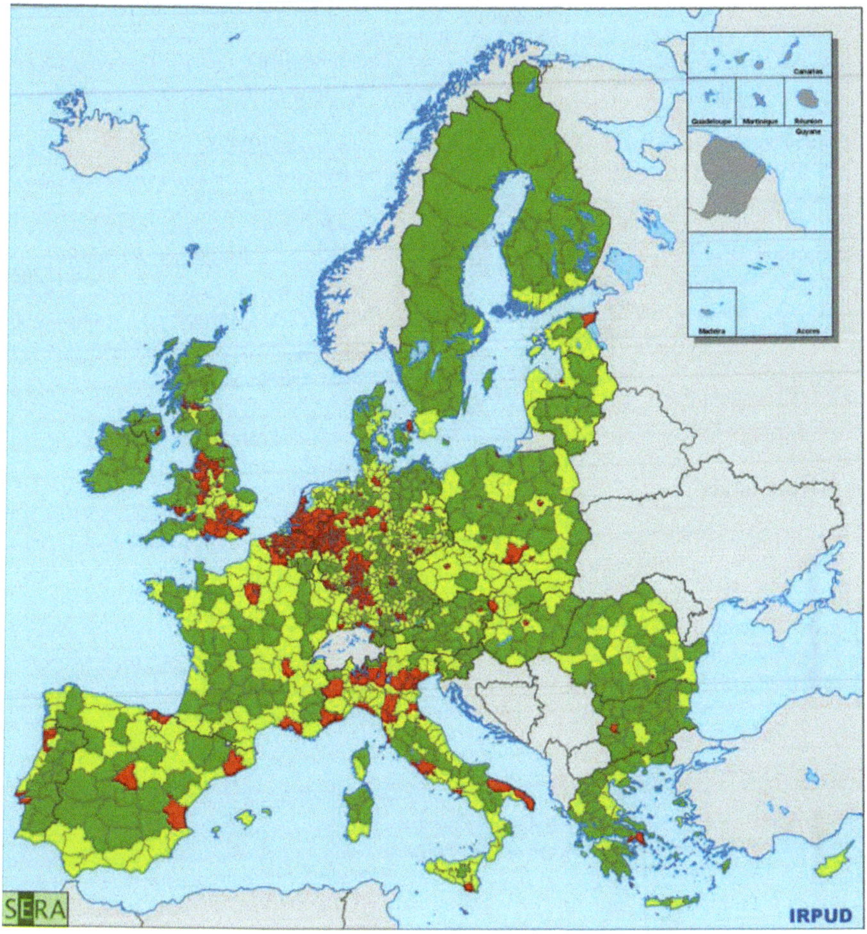

Fig. 3.8 Rural and urban regions in the EU

issues; the necessity to assess spatial and temporal information to describe a particular phenomenon, and the value of the bird's eye view on economy–environment interactions.

One might have the impression that this chapter gives no answers, but merely offers a great deal of information and data. This has been done partly deliberately, for no theory of spatio-temporal development for sustainability has yet been created and will need to be written. Perhaps these charts will help someone to do that!

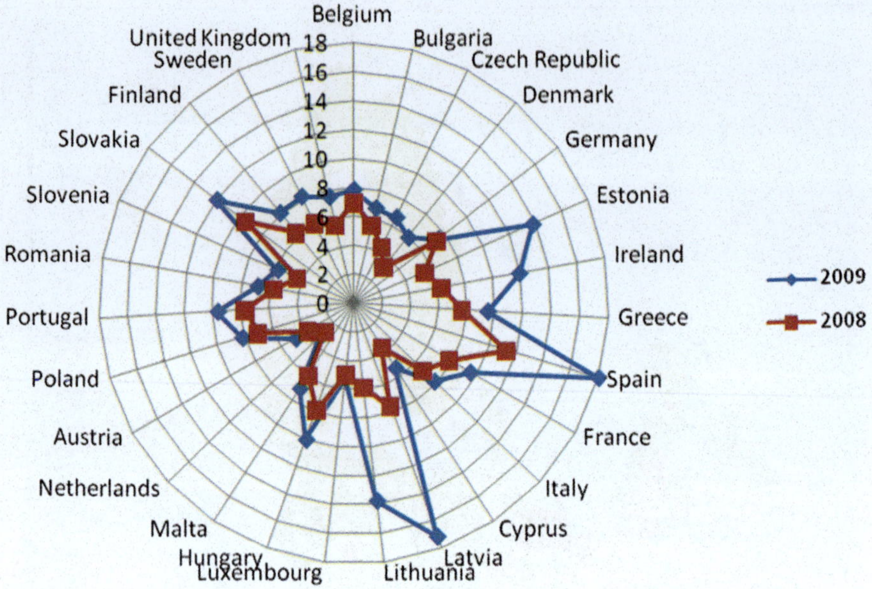

Fig. 3.9 Unemployment in the EU, 2008 and 2009

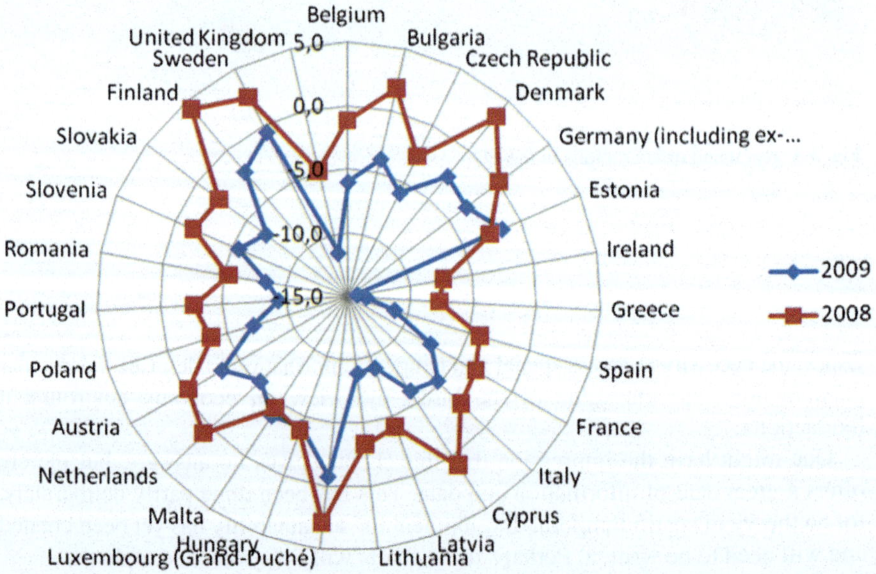

Fig. 3.10 Budget deficit in the EU countries, 2008 and 2009

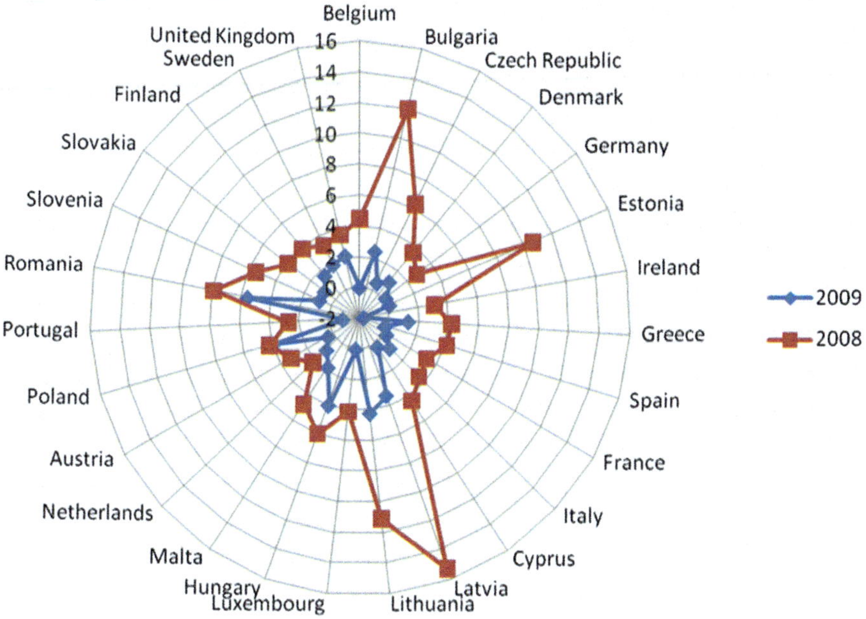

Fig. 3.11 Inflation in the EU, 2008 and 2009

References

Adler M (1998) Incommensurability and Cost-Benefit Analysis *University of Pennsylvania Law Review* 146.5 1371–1418

Aldred J (1994) Existence Value, Welfare and Altruism *Environmental Values* 3.4 381–402

Aldred J (2002) Cost-Benefit Analysis, Incommensurability and Rough Equality *Environmental Values* 11 27–47

Aldred J (2006) Incommensurability and Monetary Valuation *Land Economics* 82.2 141–161

Arrow KJ (1997) Invaluable Goods *Journal of Economic Literature* 35.2 757–765

Attfield R (1998) Existence Value and Intrinsic Value *Ecological Economics* 24.2-3 163–168

Beckerman W (1994) Sustainable Development: Is It a Useful Concept? *Environmental Values* 3.3 191–209

Carson R (1962) *Silent Spring* Boston, Hougton Mifflin Company

Common M and Perrings C (1992) Towards an Ecological Economics of Sustainability *Ecological Economics* 6.1 7–34

Daly HE (1990a) Sustainable Development: From Concept and Theory to Operational Principles *Population and Development Review* 16 25–43

Daly HE (1990b) Toward Some Operational Principles of Sustainable Development *Ecological Economics* 2.1 1–6

Dokuchaev VV (1879) *Kartografija russkich pochv. Objasnitelnyj teskst k pochvennoj karte Evropejskoj Rossii* St Petersburg, Tipografia Kirschbauma

Gallopin GC (2001) The Latin American World Model (a.k.a. the Bariloche Model): Three Decades Ago *Futures* 33.1 77–89

Goodland R and Daly H (1996) Environmental Sustainability: Universal and Non-Negotiable *Ecological Applications* 6.4 1002–1017

Griffin J (1977) Are There Incommensurable Values? *Philosophy and Public Affairs* 7.1 39–59

Holland A (1994) Values and Preferences in Environmental Economics *Environmental Values* 3.4 283–402

Kleidon A (2010) Non-Equilibrium Thermodynamics, Maximum Entropy Production and Earth-System Evolution *Philosophical Transactions of the Royal Society A: Mathematical Physical and Engineering Sciences* 368.1910 181–196

Koehler J et al (2006) Combining Energy Technology Dynamics and Macroeconometrics: The E3MG Model *Energy Journal* 27 113–133

Lapo AV (2001) Vladimir I. Vernadsky (1863–1945), Founder of the Biosphere Concept *International Microbiology* 4 47–49

Leontief W (1974) Structure of the World Economy. Outline of a Simple Input-Output Formulation *The Swedish Journal of Economics* 76.4 387–401

Leontief W (1977a) Natural Resources, Environmental Disruption, and the Future World Economy *Journal of International Affairs* 31.2 267

Leontief W (1977b) *The Future of the World Economy: A United Nations Study* New York, Oxford University Press

Leopold A (1949) *A Sand County Almanac, and Sketches Here and There* London/New York, Oxford University Press

Lovelock JE (1972) Gaia as Seen Through the Atmosphere *Atmospheric Environment (1967)* 6.8 579–580

Lovelock J (1979) *Gaia, a New Look at Life on Earth*, Oxford/New York, Oxford University Press

Lovelock JE (1990) Hands Up for the Gaia Hypothesis *Nature* 344.6262 100–102

Lovelock JE and Whitfield M (1982) Life Span of the Biosphere *Nature* 296.5857 561–563

Lutz C (2010) How to Increase Global Resource Productivity? Findings From Modelling in the PetrE Project *International Economics and Economic Policy* 7.2 343–356

Lutz C and Meyer B (2009) Economic Impacts of Higher Oil and Gas Prices: The Role of International Trade for Germany *Energy Economics* 31.6 882–887

Margulis L and Lovelock JE (1978) The Biota as Ancient and Modern Modulator of the Earth's Atmosphere *Pure and Applied Geophysics* 116.2 239–243

Martinez-Alier J, Munda G and O'Neill J (1998) Weak Comparability of Values as a Foundation for Ecological Economics *Ecological Economics* 26.3 277–286

Meadows D (1982) *Groping in the Dark: The First Decade of Global Modelling* Chichester [West Sussex]/New York, Wiley

Meadows D (1992) *Beyond the Limits: Global Collapse or a Sustainable Future* London, Earthscan Publications

Meadows D (2005) *Limits to Growth: The 30-Year Update* [3rd revision, expanded and updated edition] London, Earthscan

Meadows D and Club of Rome (1972) *The Limits to Growth a Report for the Club of Rome's Project on the Predicament of Mankind* New York, Universe Books

Mendeleev DI (1869) *Osnovy chimii* St Petersburg, Tipografija tovarischestva «Obschestvennaja polza»

Mendeleev DI (1871) *Osnovy chimii* St Petersburg, Tipografija tovarischestva «Obschestvennaja polza»

Moiseev NN (1978) Cybernetic Description of Ecological-Economic Systems *Cybernetics* 13.6 930–944

Moiseev NN (1982) *Chelovek, Sreda, Obschestvo: Problemy Formalizovannogo Opisaniya* Moscow, Nauka

Moiseev NN (1987a) Ecologicheskiy imperativ *Mir nauki* 3.03 278–284

Moiseev NN (1987b) *Man, Nature and the Future of Civilization ("Nuclea Winter" and the Problem of a "Permissible Threshold")* Moscow [accessed 18 Mar 2010] Available at http://www.ras.ru/win/db/show_ref.asp?P=.id-25085.ln-en

Moiseev NN (1988) *Ecologia chelovechestva glazami matematika: chelovek, priroda i buduschee tsivilizatsii* Moscow, Molodaya gvardia

Moiseev NN (1994) Biota kak regulator i problema Sustainability *Zhurnal vychislitelnoj matematiki i matematicheskoj fiziki* 34.4 533–544

Moiseev et al (1983) Global Models, the Biospheric Approach (Theory of the Noosphere) [accessed 18 Mar 2010] Available at http://www.iiasa.ac.at/Admin/PUB/Documents/CP-83-033.pdf

Moiseev NN, Alexandrov VV and Tarko AM (1985) *Chelovek i biosfera. Opyt sistemnogo analiza i eksperimenty s modelyami* Moscow, Nauka [accessed 18 Mar 2010] Available at http://www.ras.ru/win/db/show_ref.asp?P=.id-25414.ln-en

Naess A (1973) Shallow and Deep, Long-Range Ecology Movement. A Summary *INQUIRY An Interdisciplinary Journal of Philosophy* 16.1 95–100

Naess A (1993) *Ecology, Community and Lifestyle: Outline of an Ecosphy* (Reprinted) Cambridge, Cambridge University Press

Næss A (2008) *Ecology of Wisdom: Writings by Arne Naess* Berkeley, Counterpoint; Distributed by Publishers Group West

Norton BG (1990) Context and Hierarchy in Aldo Leopold's Theory of Environmental Management *Ecological Economics* 2.2 119–127

O'Neill J, Holland A and Light A (2007) *Environmental Values* London, Routledge

Onishi A (1977) Using a Multi-Nation Economic Model: Projection of Economic Relations Between Japan and Developing Countries in Asia (1975-1985) *Technological Forecasting and Social Change* 10.2 121–142

Onishi A (2001) The World Economy to 2015: Policy Simulations on Sustainable Development *Journal of Policy Modeling* 23.2 217–234

Onishi A (2005) Futures of Global Interdependence (FUGI) Global Modeling System: Integrated Global Model for Sustainable Development *Journal of Policy Modeling* 27.1 101–135

Onishi A (2009) A New Challenge to Economic Science: Global Model Simulation *Journal of Policy Modeling* 32.1 1–46

Pearce D (1992) Green Economics *Environmental Values* 1.1 3–13

Pezzey J (1992) Sustainability: An Interdisciplinary Guide *Environmental Values* 1.4 321–362

Randers J and Meadows D (1972) The Carrying Capacity of the Globe *Sloan Management Review (pre-1986)* 13.2 11

Roberts PC (1977) SARUM 76 – A Global Modelling Project *Futures* 9.1 3–16

Samson PR and Pitt D (1999) *Biosphere and Noosphere Reader. Global Environment, Society and Change* London, Routledge

Scolnik H (1979) A Critical Review of Some Global Models in *Global and Large Scale System Models* Lecture Notes in Control and Information Sciences Berlin/Heidelberg, Springer 58-80 Available at http://dx.doi.org/10.1007/BFb0049022

Shmeleva IA (2006) *Psychologija Ekologicheskogo Soznanija* St Petersburg, University Press

Spash CL (2007) The Economics of Climate Change Impacts À La Stern: Novel and Nuanced or Rhetorically Restricted? *Ecological Economics* 63.4 706–713

Spash CL (2008) How Much Is That Ecosystem in the Window? The One with the Bio-Diverse Trail *Environmental Values* 17 259–284

Vatn A (2000) The Environment as a Commodity *Environmental Values* 9 493–509

Vernadsky V (1924) *La Géochimie* Paris, Félix Alcan

Vernadsky VI (1929) *La Biosphere* Paris, Librairie Félix Alcan

Vernadsky VI (1936) Nauchnaya mysl kak planetarnoe javlenije

Vernadsky V (1986) *The biosphere* Oracle, Synergetic Press

Worden L (2010) Notes From the Greenhouse World: A Study in Coevolution, Planetary Sustainability, and Community Structure *Ecological Economics* 69.4 762–769

Chapter 4
Economic Valuation and Decision Making: MCDA as a Tool for the Future

Abstract Over the course of the past 50 years decisions on major infrastructure projects or projects that involve compromises between quality of environment and economic gain have been selected on the basis of cost-benefit analysis. This type of thinking masked the complexity of impact by the concept of willingness to pay and related techniques. The ecological-economic community, realising that these approaches violated the incommensurability of values principle, proposed to approach the problems of choice with the help of multicriteria decision aid tools. Over the course of several decades several tools have been developed and it is often difficult to make the right choice. This chapter presents the taxonomy of MCDA tools first proposed by Guitoni and Martel and offers several approaches to the reader. To study this subject in depth the interested reader is referred to the specialist literature, which is extensively cited in this chapter.

Keywords Multicriteria decision aid • Decision making • Taxonomy of methods • Method selection • European School

History of Multicriteria Analysis

The field of multi-criteria decision aiding (MCDA) has been developing since the 1960s. Methodological work in this field, focused on discreet methods, has been done by Roy (Roy 1985, 1991, Roy and Vincke 1981), who pioneered multi-criteria assessment with the ELECTRE family of methods, Brans (Brans et al. 1986), who created the PROMETHEE method; Hinloopen and Nijkamp, who developed the REGIME method (Hinloopen and Nijkamp 1990); Bouyssou who contributed towards the development of a concept of compensation in MCDA (Bouyssou 1986, Roy and Bouyssou 1993); Larichev, who worked on qualitative methods and the perception of multicriteria problems by a decision maker (Larichev 1987, 1996, 2001, Larichev and Brown 2000, Larichev and Moshkovich 1988, 1994, 1995, Larichev et al. 2002), Janssen who developed a DEFINITE package (Janssen 1993);

S.E. Shmelev, *Ecological Economics: Sustainability in Practice*,
DOI 10.1007/978-94-007-1972-9_4,

Bana e Costa, who developed the MACBETH method: (Costa 2001, Costa et al. 1997, 1999, 2001); Hovanov, who designed the randomised preference based method called ASPID (Afgan et al. 2000; Hovanov 1996, 2006, Hovanov et al. 2009); Munda, who developed the NAIADE method (Munda 1995, 2005a, b, Munda and Nardo 2008). The most comprehensive survey of the multi-criteria analysis methods is presented in (Figueira et al. 2005). Applications of MCDA exist for regional problems, e.g. industrial development (Nijkamp and Delft 1977), waste management (Shmelev 2003, Shmelev and Powell 2006) or renewable energy (Gamboa and Munda 2007, Madlener and Stagl 2005), environmental policy issues in Germany (Omann 2000), the Netherlands (Janssen 2001), Norway (Wenstøp and Seip 2001) and Austria (Gamper and Turcanu 2007) as well as sustainability assessment on the macro scale (Shmelev 2010a, Shmelev and Rodriguez-Labajos 2009) or macroeconomic policy (Shmelev 2010b).

MCDA Paradigm

The perspective of the MCDA presents a new paradigm, which is different from the classical goal of finding an optimal solution subject to a set of constraints characteristic of operations research, the MCDA methodology also provides an alternative to the cost-benefit analysis, the tool that was popular in the 1970s and 1980s. Within the new paradigm, a search for a compromise solution, satisfying to the decision maker, rather than the optimum, became the primary purpose of analysis (Guitouni and Martel 1998). According to Roy (2005), the choice of a monocriterion approach might:

- lead to wrongly neglecting certain aspects of realism;
- facilitate the setting up of equivalencies, the fictitious nature of which remains invisible;
- tend to present features of one particular value-system as objective.
- On the contrary, a multi-criteria approach contributes to avoiding such dangers by:
- delimiting a broad spectrum of points of view likely to structure the decision process with regard to the actors involved;
- constructing a family of criteria which preserves, for each of them, without any fictitious conversion, the original concrete meaning of the corresponding evaluations;
- facilitating debate on the respective role (weight, veto, aspiration level, rejection level, etc) that each criterion might be called upon to play during the decision aiding process.

A discrete multi-criteria problem can be described as a problem of evaluation of a finite set of alternatives according to the set of criteria, which can be expressed in the quantitative or qualitative form (Munda 1995). The MCDA methodological procedure can be seen as a non-linear recursive process composed of four steps

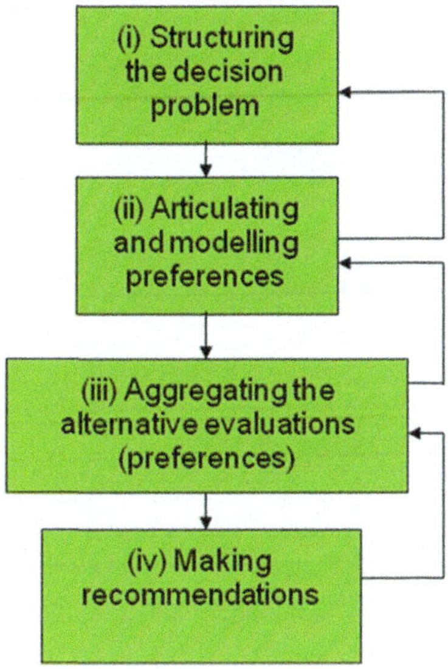

Fig. 4.1 Multicriteria decision making steps: a recursive framework

(Guitouni and Martel 1998): (i) structuring of the decision problem, (ii) articulating and modelling preferences, (iii) aggregating the alternative evaluations (preferences) and (iv) making recommendations.

Roy (2005) identifies the following basic steps in the MCDA procedure: (i) identification of alternatives; (ii) selection of the family of criteria; (iii) the choice of the problematique, the latter could be reformulated as clarification of the type of problem, the form of results, and the selection of the most appropriate procedure to guide the investigation.

According to Roy (2005) most frequently used decision aiding methods are based on mathematically explicit multi-criteria aggregation procedures (MCAP). By definition, an MCAP is a procedure, which, for any pair of potential actions, gives a clear answer to the aggregation problem. This implies:

1. various inter-criteria parameters, such as weights, scaling constants, veto, aspiration levels, rejection levels, which allow us to define the specific role that each criterion can play with respect to others;
2. a logic of aggregation, which usually takes into account: the possible types of dependence which we might want to bring into play concerning criteria;
3. the conditions under which we accept or refuse compensation between "good" and "bad" performances.

Roy emphasises the significance of the logic of aggregation of the MCAP considered. He differentiates two types of MCAP operational approaches: an approach based on a synthesising criterion and one based on a synthesising preference relational system (Fig. 4.1).

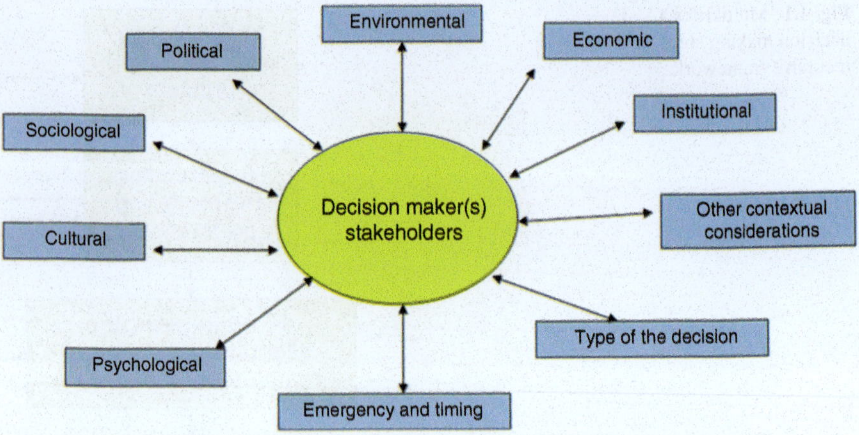

Fig. 4.2 Factors to be taken into account when making decisions: structuring the decision problem

Table 4.1 Types of scales

	Scale	Definition	Mathematical structure	Example
1	Nominal	Categorical data	Unordered set	Types of polluting gases
2	Ordinal	Describe order, but not relative size or degree of difference between the items measured	Ordered set	Environmental quality as "bad", "moderate" and "good"
3	Interval	Differences could be measured and assessed	Affine set	Temperature
4	Ratio	Estimation of the ratio between a magnitude of a continuous quantity and a unit magnitude of the same kind	Field	Mass, time, energy

According to Einstein "The formulation of a problem is often more essential than its solution, which may be merely a matter of mathematical or experimental skill" (Fig. 4.2).

Three basic concepts play a fundamental role in MCDA (Roy 1996) those of an alternative, criterion and problematique:

Alternative (or potential action) a – the object of decision:

A set of alternatives $A = \{a_1, a_2, \ldots, a_m\}$, when the number of actions is finite;

Criterion g is a tool constructed for evaluating and comparing potential actions according to a point of view which must be well-defined. $g(a)$ is called the performance of a according to this criterion.

Often it is necessary to define all possible evaluations to which a criterion can lead. The following types of scales (Stevens 1946) (Table 4.1)

Scales of 2, 3, and 4 are all could be used in multicriteria decision aid: ordinal, interval and ratio scale.

Types of MCDA Problematic

According to Prof. Bernard Roy (1996), the Decision Making Situation can be categorised according to some decision problematic:

Description problematic (P, δ): the aid helps to answer the following questions: In what terms should we pose the problem? What type of results should we try to obtain? How does the analyst see himself fitting into the decision process to aid in arriving at these results? What kind of procedure seems the most appropriate for guiding his/her investigation?

The choice problematic (P, α): the aid is oriented towards a selection of a small number (as small as possible) of "good" actions in such a way that a single alternative or a subset may finally be chosen; the subset N of the selected actions could contain all the most satisfying actions, which remain non comparable between one another;

The sorting problematic (P, β): the aid is oriented towards an assignment of each action to one category (judged the most appropriate) among those of a family of predefined categories: e.g. the family of four categories could contain: (i) actions for which implementation is fully justified; (ii) could be advised after only minor modifications, (iii) can only be advised after major modifications; (iv) is inadvisable;

The ranking problematic (P, γ): the aid is oriented towards creation of a complete or partial preorder on A, which can be regarded as an appropriate instrument for comparing actions between one another;

These four major approaches to the multicriteria decision aid represent the framework that helps to find which method would work best in each concrete case. The guidelines for selecting the best method will be explored later in this chapter.

If we use the notation accepted in this chapter and assume that:

$A = \{a_1,\dots,a_i,\dots, a_m\}$ – alternatives
$G = \{g_1,\dots,g_j,\dots g_n\}$ – criteria

Then E = Evaluation matrix will take the form of that exhibited in Table 4.2.

Table 4.2 Evaluation matrix E for multicriteria analysis for road building

		g_1	…	g_j	…	g_n
		Costs		Forest lost		Effect of emissions on public health
a_1	Straight through the nature reserve	e_{11}	…	e_{1j}	…	e_{1n}
…		…	…	…	…	…
a_i	Through major cities	e_{i1}	…	e_{ij}	…	e_{in}
…		…	…	…	…	…
a_m	Avoiding both nature reserve and cities	e_{m1}	…	e_{mj}	…	e_{mn}

As an element of innovative change contrasting with neoclassical microeconomic theory Prof. Bernard Roy (1985) introduced the following four elementary binary relations:

1. *a* **I** *b (indifference situation: a* is indifferent to *b;*
2. *a* **P** *b (preference situation): a* is strictly preferred to *b;*
3. *a* **Q** *b (weak preference situation):* is the hesitation between the indifference and preference situations and not being sure that *(a P b);*
4. *a* **R** *b (incomparability situation):* in this situation the hesitation is between *a* **P** *b and b* **P** *a*

In this context the weak preference situation and the incomparability situation are two elements that distinguish this approach from the others and allow us to capture more nuances and allow adequate representation of the real life choices.

Selecting the Right Method

But which of the dozens of methods available should be the best for a particular practical situation? The researchers Martel and Guitoni (Guitouni and Martel 1998) propose an interesting approach to the selection of a MCDA method:

Guideline G1: Determine the stakeholders of the decision process. If there are many decision makers (judges), one should think about group decision making methods or group decision support systems (GDSS).

Guideline G2: Consider DM "cognition" (DM way of thinking) when choosing a particular preference evaluation mode. If he is more comfortable with pairwise comparisons, why use tradeoffs and vice versa?

Guideline G3: Determine the decision problematic pursued by the DM. If the DM wants to achieve an alternatives ranking, then a ranking method is appropriate

Guideline G4: Choose the MCAP that can handle properly the input information available and for which the DM can easily provide the required information; the quality and quantity of information are major factors in the choice of the method).

Guideline G5: The compensation degree of the MCAP method is an important aspect to consider and to explain to the DM. If he refuses any compensation, then many MCAP will not be considered.

Guideline G6: The fundamental hypotheses of the method are to be met (verified), otherwise one should choose another method.

Guideline G7: The decision support system implicit in the method is an important aspect to be considered when the time comes to choose a MCDA method.

Figures 4.3–4.5 outline the method selection procedure in diagrammatic language.

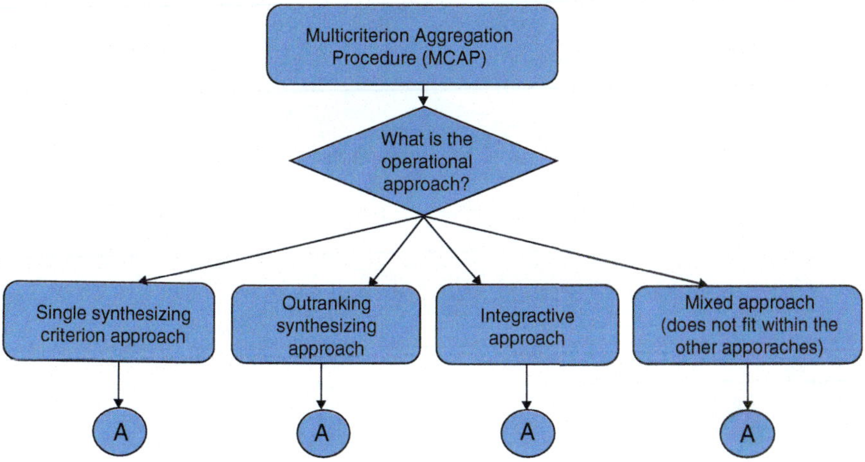

Fig. 4.3 First stage of the method selection procedure

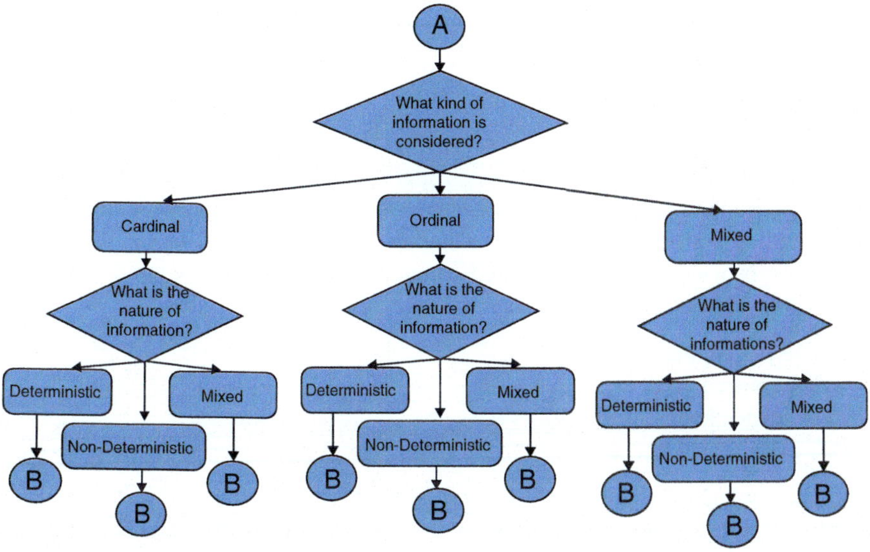

Fig. 4.4 Second stage of the method selection procedure

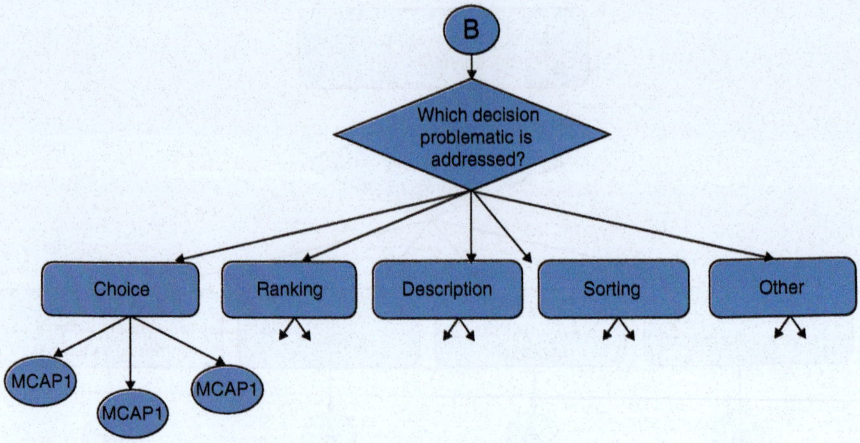

Fig. 4.5 Third stage of the method selection procedure

Tables 4.3–4.8 present the main structural characteristics of the MCDA tools as identified by Guitoni and Martel.

Sustainability Assessment with MCDA

It would be most appropriate here to consider several examples of the application of multicriteria methods to the analysis of sustainability on the macro scale. We will explore these issues in more depth in Chap. 7.

Analysis of sustainability of the UK economy shown in Fig. 4.6 has been carried out with the aid of the NAIADE method using three criteria

- GDP per capita;
- CO_2 emissions;
- Life expectancy (Table 4.9).

The year 2002 (H) dominates all other years, and the next year in line is 2005 (K), after which 1995 (A), 1997(C) and 2003(I) follow, followed by 2004 (J), and then 1998(D) and 2001 (G), with the bottom place occupied by 2000 (F), 1999 (E) and 1996 (B).

Table 4.3 Taxonomy of MCDA methods (Source: Guitouni and Martel 1998): Elementary methods

	Guideline									
	G2				G3	G4				
	Preference				Decision	Kind of information			Information features	
Method	eluc.mode	Moment	Pref. structure	Order	probl.	Ord.	Card.	Mix.	Deter.	Non deter.
Elementary methods										
Weighted sum	Direct rating	a priori	{**P, I**}	Total preorder	α		✓		✓	
Lexicographic method	Direct rating	a priori	{**P, I**}	Total preorder	α	✓	✓	✓	✓	
Conjunctive method	Direct rating	a priori	{**P, I**}	Filtration	ν/α	✓	✓	✓	✓	
Disjunctive method	Direct rating	a priori	{**P, I**}	Filtration	ν/α	✓	✓	✓	✓	
Maximin method	Direct rating	a priori	{**P, I**}	Total preorder	α	✓	✓		✓	

Table 4.4 Taxonomy of MCDA methods (Source: Guitouni and Martel 1998): Single synthesising criterion

	Guideline									
	G2				G3	G4				
						Kind of information			Information features	
Method	Preference eluc.mode	Moment	Pref. structure	Order	Decision probl.	Ord.	Card.	Mix.	Deter.	Non deter.
Single synthesising criterion										
Fuzzy weighted sum	Direct rating	a priori	{**P, I**}	Semi-order	α	✓	✓	✓		✓
TOPSIS	Direct rating	a priori	{**P, I**}	Total preorder	α		✓		✓	
MAVT	Tradeoffs	a priori	{**P, I**}	Total preorder	α		✓		✓	
UTA	Tradeoffs	a priori	{**P, I**}	Total preorder	α	✓			✓	
SMART	Tradeoffs & rating	a priori	{**P, Q, I**}	Total preorder	α		✓		✓	
MAUT	Tradeoffs & lotteries	a priori	{**P, I**}	Total preorder	α		✓			✓
AHP	Pairwise comparison	a priori	{**P, I**}	Total preorder	α, γ		✓		✓	✓
EVAMIX	Direct rating	a priori	{**P, I**}	Total preorder	α, γ	✓	✓	✓	✓	
Fuzzy maximin	Direct rating	a priori	{**P, Q, I**}	Semi-order	α	✓	✓		✓	✓

Table 4.5 Taxonomy of MCDA methods (Source: Guitouni and Martel 1998): Outranking methods

	Guideline									
	G2				G3	G4				
						Kind of information			Information features	
Method	Preference eluc. mode	Moment	Pref. structure	Order	Decision probl.	Ord.	Card.	Mix.	Deter.	Non deter.
Outranking methods										
ELECTRE I	Pairwise comparison	a priori	{**S, R**}	Core	α	✓	✓	✓	✓	
ELECTRE II	Pairwise comparison	a priori	{$\mathbf{S^F}$, $\mathbf{S_f}$, **R**}	Partial semi-order	γ	✓	✓	✓	✓	
ELECTRE III	Pairwise comparison	a priori	Valued {**S, R**}	Partial semi-order	γ	✓	✓	✓	✓	
ELECTRE IV	Pairwise comparison	a priori	{$\mathbf{S^1}$, $\mathbf{S^2}$, $\mathbf{S^3}$, $\mathbf{S^4}$, $\mathbf{S^5}$, **R**}	Partial preorder	γ	✓	✓	✓	✓	
ELECTRE IS	Pairwise comparison	a priori	{**S, R**}	Partial semi-order	α	✓	✓	✓	✓	
ELECTRE TRI	Pairwise comparison	a priori	{**S, R**}	Partial interval order	β	✓	✓	✓	✓	
PROMETHEE I	Pairwise comparison	a priori	Valued {**P, I, R**}	Partial semi-order	γ	✓	✓	✓	✓	
PROMETHEE II	Pairwise comparison	a priori	Valued {**P, I**}	Total preorder	γ	✓	✓	✓	✓	
MELCHIOR	Pairwise comparison	a priori	Valued {**P, I**}	Total preorder	γ	✓			✓	
ORESTE	Pairwise comparison	a priori	{**P, Q, I**}	Semi-order	γ	✓			✓	
REGIME	Pairwise comparison	a priori	{**SF, Sf, R**}	Total preorder	γ	✓			✓	
NAIADE	Pairwise comparison	a priori	Valued {**S, R**}	Total preorder	γ	✓	✓	✓	✓	✓

Table 4.6 Taxonomy of MCDA methods (Source: Guitouni and Martel 1998), Elementary methods, G5–G7

	Guideline					
	G5			G6		G7
Method	Discrimination power of the criteria	Compensation	Information inter-criteria	Hypothesis	MCAP treatment	Software package
Elementary methods						
Weighted sum	Absolute	Totally	Total and explicit importance coeff.	ind., com., inv., tran., dom.	Algebraic sum	
Lexicographic method	Absolute	Non	n/a	ind., inv., tran., dom.	Cutting planes	
Conjunctive method	Absolute	Non	n/a	ind., inv., tran., dom.	Thresholds	
Disjunctive method	Absolute	Non	n/a	ind., inv., tran., dom.	Thresholds	
Maximin method	Absolute	Non	n/a	ind., inv., tran., dom.	Max and min operators	

Table 4.7 Taxonomy of MCDA methods (Source: Guitouni and Martel 1998), Single synthesising criterion, G5–G7

	Guideline					
	G5			G6		G7
Method	Discrimination power of the criteria	Compensation	Information inter-criteria	Hypothesis	MCAP treatment	Software package
Single synthesising criterion						
Fuzzy weighted sum	Non-absolute	Totally	Total and explicit	ind., com., inv., tran., dom.	a-cut and fuzzy arithm.	
TOPSIS	Absolute	Totally	Total and explicit	ind., com., inv., tran., dom.	Euclidian distances	
MAVT	Absolute	Partially	Total and explicit	ind., inv., tran., dom.	Value aggregation (sum or mult)	✓
UTA	Absolute	Partially	Indirect	ind., inv., tran., dom.	Value aggregation (sum)	✓
SMART	Absolute	Partially	Total and explicit	ind., com., inv., tran., dom.	Value aggregation (sum)	✓
MAUT	Absolute	Partially	Total and explicit	ind., inv., tran., dom.	Utility aggregation (sum or mult)	✓
AHP	Absolute	Partially	Total and explicit	Inner and outer ind., inv., dom.	Eigenvector method	✓
EVAMIX	Absolute	Partially	Total and explicit	ind., com., inv., tran., dom.	Algebraic sum	
Fuzzy maximin	Non-absolute	Non	n/a	ind., com., inv., dom.	Max and min operators	

Table 4.8 Taxonomy of MCDA methods (Source: Guitouni and Martel 1998), Outranking methods, G5–G7

	Guideline					
	G5			G6		G7
Method	Discrimination power of the criteria	Compensation	Information inter-criteria	Hypothesis	MCAP treatment	Software package
Outranking methods						
ELECTRE I	Absolute	Partially	Total and explicit	ind., inv., coal.	Graph theory (core)	✓
ELECTRE II	Absolute	Partially	Total and explicit	ind., inv., coal.	Graph theory (distillation)	✓
ELECTRE III	Non-absolute	Partially	Total and explicit	ind., inv., coal.	Graph theory (distillation)	✓
ELECTRE IV	Non-absolute	Partially	n/a	ind., inv., coal.	Graph theory (distillation)	✓
ELECTRE IS	Non-absolute	Partially	Total and explicit	ind., inv., coal.	Graph theory (core)	✓
ELECTRE TRI	Non-absolute	Partially	Total and explicit	ind., inv., coal.	Disjunctive and conjunctive	✓
PROMETHEE I	Non-absolute	Partially	Total and explicit	ind., inv., coal.	Leaving and entering flows	✓
PROMETHEE II	Non-absolute	Partially	Total and explicit	ind., inv., coal.	Leaving and entering flows	✓
MELCHIOR	Non-absolute	Partially	Total order	ind., inv.	Graph theory (distillation)	
ORESTE	Absolute	Partially	Total preorder	ind., inv., coal.	Graph theory	✓
REGIME	Absolute	Partially	Total order	ind., inv.	Graph theory	✓
NAIADE	Non-absolute	Partially	n/a	ind., inv.	Fuzzy arithm and leaving and entering flows	✓

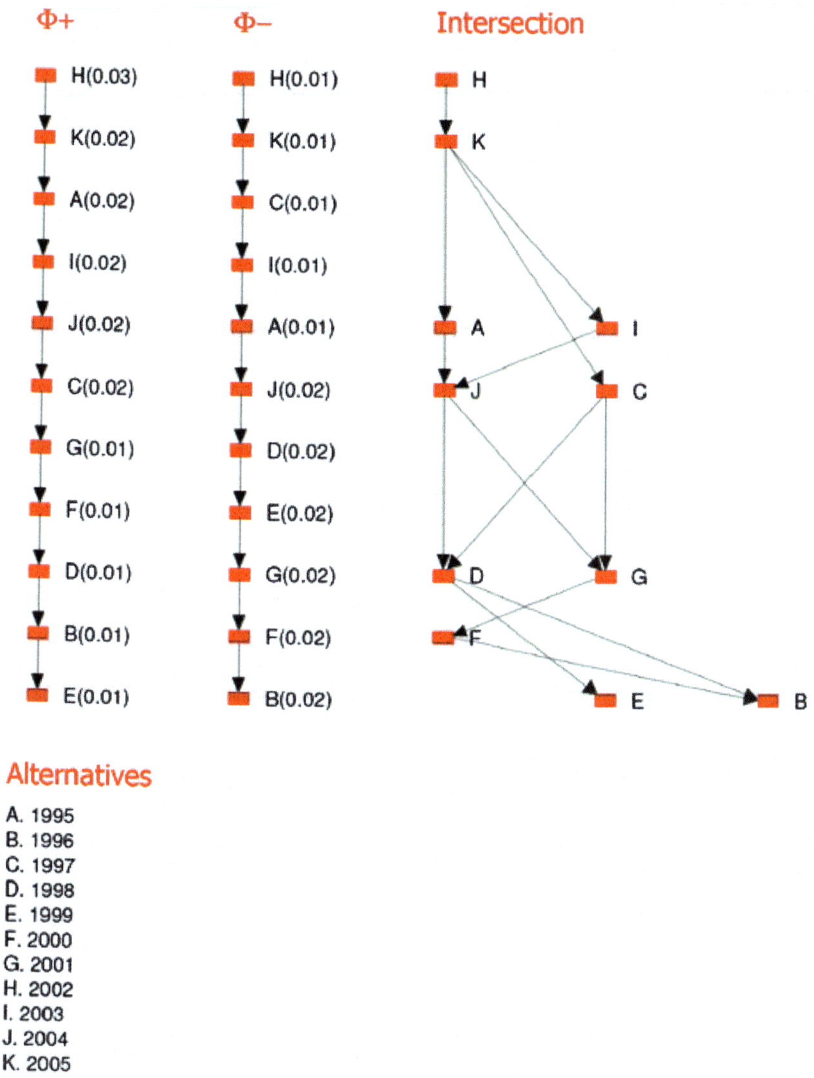

Fig. 4.6 The Application of MCDA to UK strong sustainability analysis (1995–2005)

Table 4.9 Evaluation matrix E for multicriteria analysis for sustainability assessment

		g_1	g_j	g_n
		GDP per capita	CO_2 emissions	Life expectancy
a_1	1995			
a_2	1996			
a_3	1997			
a_4	1998			
a_5	1999			
a_6	2000			
a_7	2001			
a_8	2002			
a_9	2003			
a_{10}	2004			
a_{11}	2005			

References

Afgan NH, Carvalho MG and Hovanov NV (2000) Energy System Assessment with Sustainability Indicators *Energy Policy* 28.9 603–612

Bouyssou D (1986) Some Remarks on the Notion of Compensation in MCDM *European Journal of Operational Research* 26.1 150–160

Brans JP, Vincke P and Mareschal B (1986) How to Select and How to Rank Projects: The Method *European Journal of Operational Research* 24.2 228–238

Costa CABE (2001) The Use of Multi-Criteria Decision Analysis to Support the Search for Less Conflicting Policy Options in a Multi-Actor Context: Case Study *Journal of Multi-Criteria Decision Analysis* 10.2 111–125

Costa CABE, Stewart TJ and Vansnick J-C (1997) Multicriteria Decision Analysis: Some Thoughts Based on the Tutorial and Discussion Sessions of the ESIGMA Meetings *European Journal of Operational Research* 99.1 28–37

Costa CABE et al (1999) Decision Support Systems in Action: Integrated Application in a Multicriteria Decision Aid Process *European Journal of Operational Research* 113.2 315–335

Costa CABE, Silva FND and Vansnick J-C (2001) Conflict Dissolution in the Public Sector: A Case-Study *European Journal of Operational Research* 130.2 388–401

Figueira J, Greco S and Ehrgott M (2005) *Multiple-Criteria Decision Analysis. State of the Art Surveys* New York, Springer

Gamboa G and Munda G (2007) The Problem of Windfarm Location: A Social Multi-Criteria Evaluation Framework *Energy Policy* 35.3 1564–1583

Gamper CD and Turcanu C (2007) On the Governmental Use of Multi-Criteria Analysis *Ecological Economics* 62.2 298–307

Guitouni A and Martel J-M (1998) Tentative Guidelines to Help Choosing an Appropriate MCDA Method *European Journal of Operational Research* 109.2 501–521

Hinloopen E and Nijkamp P (1990) Qualitative Multiple Criteria Choice Analysis *Quality and Quantity* 24.1 37–56

Hovanov NV (1996) *Analiz i sintez pokazatelej pri informazionnom defizite.* St Petersburg University Press

Hovanov N (2006) Multicriteria Estimation of Probabilities on Basis of Expert Non-numeric, Non-exact and Non-complete Knowledge *MCDM 2006 June 19–23 2006* 8

Hovanov N, Yudaeva M and Hovanov K (2009) Multicriteria Estimation of Probabilities on Basis of Expert Non-numeric, Non-exact and Non-complete Knowledge *European Journal of Operational Research* 195.3 857–863

Janssen R (1993) *Multiobjective Decision Support for Environmental Management* Dordrecht, Kluwer Academic Publishers

Janssen R (2001) On the Use of Multi-Criteria Analysis in Environmental Impact Assessment in the Netherlands *Journal of Multi-Criteria Decision Analysis* 10.2 101–109

Larichev OI (1987) *Objektivnye Modeli I Subjektivnye Reshenija* Moscow, Nauka

Larichev OI (1996) *Kachestvennye Metody Prinyatija Reshenij. Verbalnyj Analys Reshenij* Moscow, Nauka

Larichev OI (2001) Ranking Multicriteria Alternatives: The Method ZAPROS III *European Journal of Operational Research* 131.3 550–558

Larichev OI and Brown RV (2000) Numerical and Verbal Decision Analysis: Comparison on Practical Cases *Journal of Multi-Criteria Decision Analysis* 9 263–273

Larichev OI and Moshkovich HM (1988) Limits to Decision-Making Ability in Direct Multiattribute Alternative Evaluation *Organizational Behavior and Human Decision Processes* 42.2 217–233

Larichev OI and Moshkovich HM (1994) An Approach to Ordinal Classification Problems *International Transactions in Operational Research* 1.3 375–385

Larichev OI and Moshkovich HM (1995) ZAPROS-LM – A Method and System for Ordering Multiattribute Alternatives *European Journal of Operational Research* 82.3 503–521

Larichev OI, Kortnev AV and Kochin DY (2002) Decision Support System for Classification of a Finite Set of Multicriteria Alternatives *Decision Support Systems* 33.1 13–21

Madlener R and Stagl S (2005) Sustainability-Guided Promotion of Renewable Electricity Generation *Ecological Economics* 53.2 147–167

Munda G (1995) *Multicriteria Evaluation in a Fuzzy Environment* Heidelberg, Physica-Verlag

Munda G (2005a) "Measuring Sustainability": A Multi-Criterion Framework *Environment Development and Sustainability* 7.1 117–134

Munda G (2005b) Multiple Criteria Decision Analysis and Sustainable Development *in Multiple-Criteria Decision Analysis. State of the Art Surveys* New York, Springer 953–986

Munda G and Nardo M (2008) Noncompensatory/Nonlinear Composite Indicators for Ranking Countries: A Defensible Setting. *Applied Economics* 99999.1 1

Nijkamp P and Delft A (1977) *Multicriteria Analysis and Regional Decision Making* New York, Springer

Omann I (2000) How Can Multi-Criteria Decision Analysis Contribute to Environmental Policy Making? A Case Study on Macro-Sustainability in Germany *in 3rd International Conference of the European Society for Ecological Economics, May 3–6, 2000* Vienna 26

Roy B (1985) *Methodologie Multicritere d'aide a la Decision* Paris, Economica

Roy B (1991) The Outranking Approach and the Foundations of Electre Methods *Theory and Decision* 31.1 49–73

Roy B (1996) *Multicriteria Methodology for Decision Aiding* Dordrecht/Boston, Kluwer Academic Publishers

Roy B (2005) Paradigms and Challenges *in* Figueira J Greco S and Ehrgott M *Multiple-Criteria Decision Analysis. State of the Art Surveys* New York, Springer 3–24

Roy B and Bouyssou D (1993) *Aide Multicritere a la Decision: Methodes et Cas* Paris, Economica

Roy B and Vincke P (1981) Multicriteria Analysis: Survey and New Directions *European Journal of Operational Research* 8.3 207–218

Shmelev SE (2003) *Ekologo-Ekonomicheskoe Modelirovanie Regionalnych System Upravlenia Otchodami (Ecological-Economic Modelling of the Regional Waste Management Systems, PhD Thesis)* St Petersburg, State University

Shmelev SE (2010a) *Dynamic Sustainability Assessment: The Case of Russia in the Period of Transition (1985–2007)* Oxford, University of Oxford

Shmelev SE (2010b) *Environmentally Extended Input-Output Analysis of the UK Economy: Key Sector Analysis* Oxford, University of Oxford
Shmelev SE and Powell JR (2006) Ecological-Economic Modelling for Strategic Regional Waste Management Systems *Ecological Economics* 59.1 115–130
Shmelev SE and Rodriguez-Labajos B (2009) Dynamic Multidimensional Assessment of Sustainability at the Macro Level: The Case of Austria *Ecological Economics* 68.10 2560–2573
Stevens SS (1946) On the Theory of Scales of Measurement *Science* 103.2684 677–680
Wenstøp F and Seip K (2001) Legitimacy and Quality of Multi-Criteria Environmental Policy Analysis: A Meta Analysis of Five MCE Studies in Norway *Journal of Multi-Criteria Decision Analysis* 10.2 53–64

Chapter 5
Macroeconomy: Market Failures and Externalities: What Can Be Done

Abstract The concept of market failure has become central in defining the place, scope and effectiveness of government intervention in case of environmental pollution. This chapter presents the classic model of externalities and explains why current economic theories have not been very successful in fully tackling them. The concept of an environmental tax, designed in the spirit of the Arthur Pigou tax is introduced. Comparative analysis of various environmental taxes in major EU countries is made. The concept of an environmental tax is illustrated with the example of a landfill tax in two EU Member States: Austria and the Netherlands. It can be seen how different rates of environmental tax and the pace of their introduction could affect the development pattern of a municipal waste management system.

Keywords Market failure • Externality • Pigou Taxes • Environmental taxes • Landfill tax

Economic Theory

The history of economic thought gives us a wealth of examples, where influential ideas proposed by one of the economists were later expanded by them to include the issue of the interrelationship between the economy and the environment. Arthur Pigou developed an idea of environmental taxation, Ronald Coase formulated several institutional solutions to the problem of social cost, Wassily Leontief suggested a way to analyse the environmental effects of macroeconomic activity. Gunnar Myrdal explored the interdependence of economic, social and institutional phenomena, Leonid Kantorovich proposed linear programming as a tool for the optimal allocation of resources, Richard Stone developed a system of national accounting, Amartia Sen worked on poverty issues, Daniel Kanemann explored the psychological deficiencies of rational behaviour and decision making and Elinor Ostrom explored issues of economic governance of the Commons. All these contributions

S.E. Shmelev, *Ecological Economics: Sustainability in Practice*,

DOI 10.1007/978-94-007-1972-9_5,

Box 5.1 Important Theoretical Contributions in Economics

- A. Marshall (1842–1924, principles of economics, 1890)
- T. Veblen (1857–1929), institutional economics
- A. Pigou (1877–1959, welfare economics, Wealth and Welfare, 1912, 1920, internalisation of externalities)
- J. M. Keynes (1883–1946), interventionist policies, business cycles, recession
- J. Hicks (1904–1989, IS-LM model, Nobel Prize, 1972)
- R. Coase (1910–), Nobel Prize, 1991; The Problem of Social Cost (the Coase Theorem) (1960)
- W. Leontief (1906–1999), Nobel Prize, 1973, input-output method
- G. Myrdal (1898–1987), Nobel Prize, 1974, interdependence of economic, social and institutional phenomena
- L. Kantorovich (1912–1986), Nobel Prize, 1975, optimum allocation of resources, linear programming
- R. Stone (1913–1991) Nobel Prize, 1984, development of the system of national accounts
- J. Nash (1928–), Nobel Prize, 1994 Noncooperative games
- Sen (1933–), Nobel Prize, 1998 welfare economics
- J. Stiglitz (1943–), Nobel Prize, 2001 Globalization and its Discontents, information theory
- D. Kahneman (1934–), Nobel Prize, 2002, psychology, uncertainty and decision making
- Elinor Ostrom (1933–) economic governance of the commons

are highly relevant for ecological economics, and many links between the writings of the above mentioned economists and the ecological economics community have already been established.

Externalities

An external effect, or an externality, is said to occur when the production or consumption decisions of one agent have an impact on the utility or profit of another agent in an unintended way, and when no compensation/payment is made by the generator of the impact to the affected party.

According to David Pearce, an externality is a detrimental (or beneficial) effect to a third party for which no price is exacted (Pearce 2002).

Starting with the pioneering works by Arthur Pigou (1920), which later became the basis for the introduction of environmental taxes in the EU and other parts of the world, and Harold Hotelling (1931), who worked on the economics of exhaustible resources, environmental and resource economics received its solid foundation.

Box 5.2 Important Environmental Economics Contributions

- Pearce D. (2002) ***An Intellectual History of Environmental Economics***, Annu. Rev. Energy Environ, 2002
- Coase R.H. (1960) ***The Problem of Social Cost***, Journal of Law and Economics, Vol. 3, pp. 1-44
- *Ayres R., Kneese A.V. (1969)* ***Production, Consumption, and Externalities***, *The American Economic Review, Vol. 59, Issue 3, pp. 282-297*
- *Leontief W. (1970)* ***Environmental Repercussions and the Economic Structure: An Input-Output Approach***, *The Review of Economics and Statistics, Vol. 52, Issue 3, pp. 262-271*
- Kapp. W. (1970) **Environmental disruption: General issues and methodological problems,** *Social Science Information 1970; 9 (4), pp. 15-32*
- Baumol W. (1972) ***On the Taxation and the Control of Externalities***, American Economic Review, Vol. 62, Issue 3, pp. 307-322
- *Kneese A. (1971)* ***Environmental Pollution: Economics and Policy***, *American Economic Review, Vol. 61, Issue 2, pp. 153-166*
- *Baumol W.J. and Oates W. E. (1971)* ***The Use of Standards and Prices for Protection of the Environment***, *The Swedish Journal of Economics, Vol. 73, Issue 1, pp. 42-53*
- *Myrdal G. (1973)* ***Economics of an improved environment***, *World Development, Vol. 1, Issue 1–2, pp. 102-114*
- *Tietenberg T. (1973)* ***Controlling Pollution by Price and Standard Systems: A General Equilibrium Analysis***, *The Swedish Journal of Economics, Vol. 75, Issue 2, pp. 193-203*

Later on due to contributions by Ronald Coase (1960), Robert Ayres and Kneese (1969), Wassily Leontief (1970), William Baumol (1972, Baumol and Oates 1971), Allen Kneese (1971), Gynnar Myrdal (1973), Tom Tietenberg (1973) the basis was created for the sound environmental policy, the application of various instruments to regulate the quality of the environment.

One can argue that ecological economics builds on the achievements of environmental economics, but goes further on a range of important issues. Let us consider the basis for the economic regulation of environmental quality.

Imagine that we are dealing with a standard economic good. The level of production of a standard economic good will be determined at the intersection of the marginal private benefit (MPB) and marginal private cost (MPC) curves and will amount to Q_1 (Fig. 5.1).

If we are dealing with a good, where an externality is present, the marginal social costs (MSC) curve is going to differ from the marginal private costs (MPC) curve by the value of marginal external costs (MEC). The solution according to Pigou is to "internalise" external costs by including them in production costs by means of a so called "Pigou tax" (Fig. 5.2). The level of production in this case will be reduced automatically.

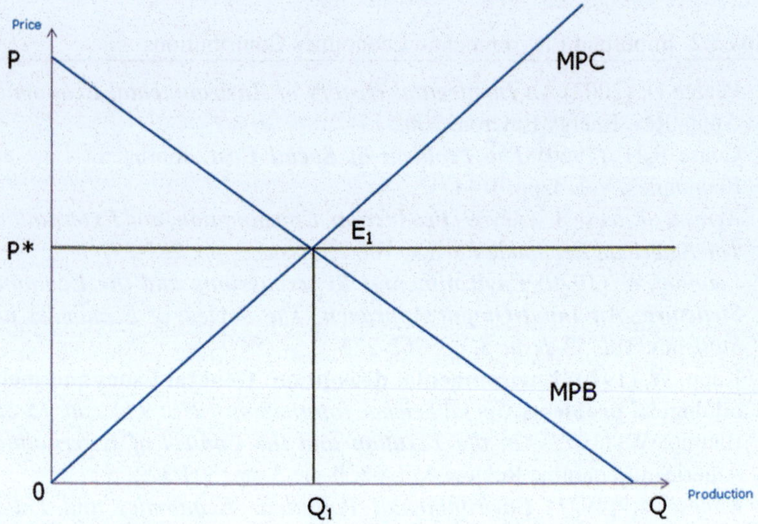

Fig. 5.1 Normal economic good

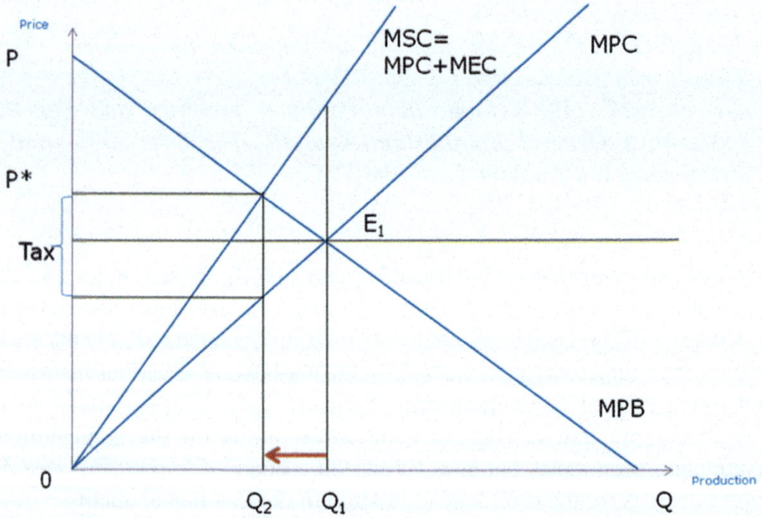

Fig. 5.2 Pigou tax

An alternative depiction of the effect of the Pigou tax is presented in Figs. 5.3 and 5.4.

Although Pigou's thinking is considered to be an important element in the theory of externalities and to essentially underpin the introduction of environmental taxes in many European countries, the theory has been criticised in the literature. The criticism comes from the fact that the situation where there is only one polluter affecting only one recipient or one aspect of environmental quality (Fig. 5.5) is extremely rare.

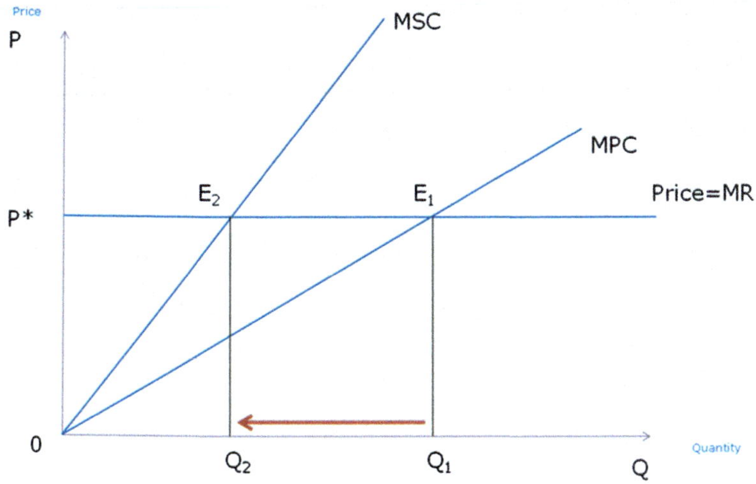

Fig. 5.3 Desirable reduction in production quantity

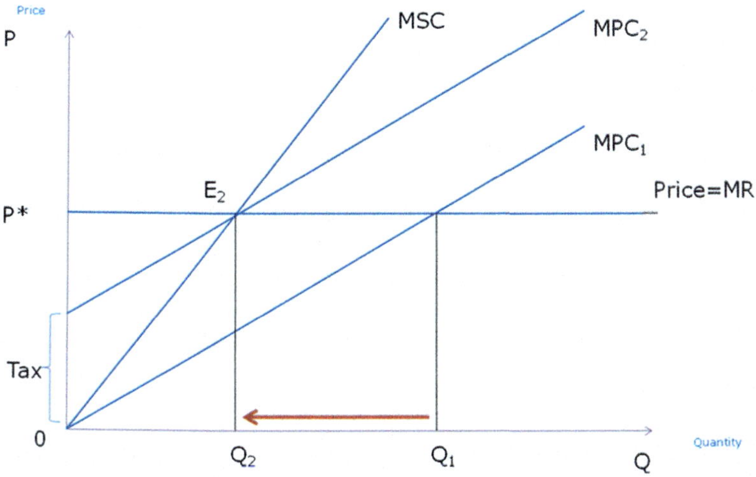

Fig. 5.4 Effect of the Pigou tax

Fig. 5.5 One polluter, one recipient case

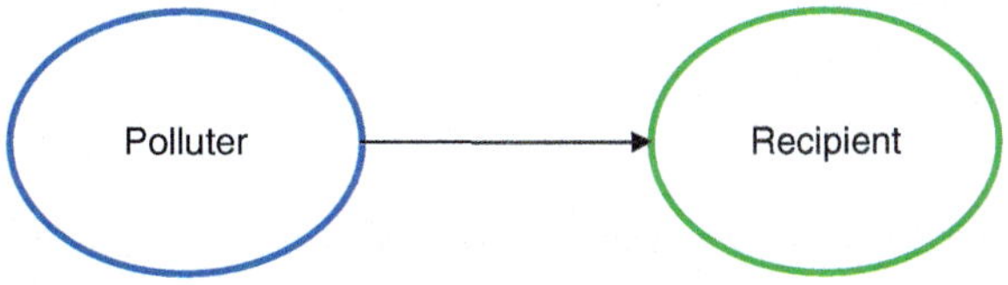

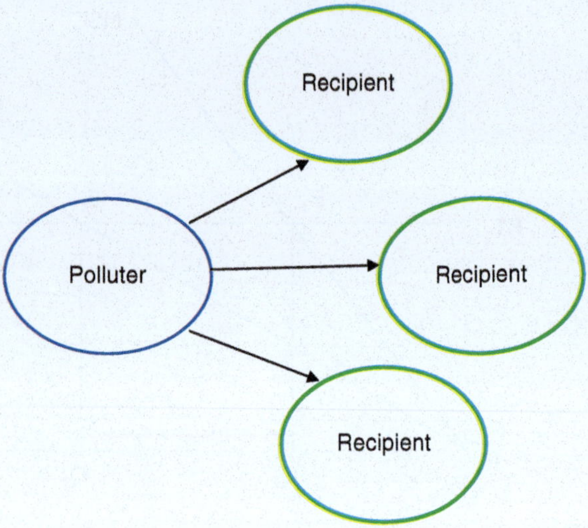

Fig. 5.6 One polluter and multiple recipients

Much more often, as in the case of sustainable waste management, sustainable energy, and ecosystem assessments, we are dealing with a situation where there are multiple recipients or several aspects of the system affected (human health, visual disamenity, valuable species, depletion of material resources, etc), Fig. 5.6. The issue of accumulation of pollution is observed, and there is also a definite time dimension for the pollution problem. In this case and in many others it becomes practically impossible to assess marginal external costs and establish the most socially desirable level of production.

In this respect, the approach where the amount of an environmental charge is set up in the iterative manner becomes an option of choice. It will be illustrated with several examples showing that the iterative approach works in certain cases, especially in waste management. Obviously, there is an important requirement for its effective operation, i.e. compliance with legislation, both national and international, which in the case of waste management means no illegal dumping and no transboundary shipment of hazardous waste.

Environmental Taxes

In this chapter we will consider environmental taxes as an example of the corrective Pigouvian instrument of environmental policy. Currently Eurostat differentiates between four categories of environmental taxes:

- Energy taxes;
- Resource taxes;
- Transport taxes;
- Pollution taxes.

Table 5.1 Environmental taxes applied in the EU countries

Country	Energy and natural gas tax	Water abstraction charges	Water pollution charges	Charge on fertilisers	Charge on pesticide	Sulphur taxes	Carbon tax	Landfill tax
Austria	1996	1892	1979	1986	No tax	No tax	No tax	1989
Denmark	1977	1994	1997	1998	1986	1996	1992	1987
Finland	1974	1996	1996	1976	No tax	1999	1990	1996
France	1969	1964	1969	No tax	2008	1985	No tax	1993
Germany	1976	1988	1981	ND	ND	ND	1999	No tax
Netherlands	1992	1995	1970	No tax	No tax	2000	1990	1995
Norway	1970	No tax	1974	1988	1988	1970	1991	1999
Sweden	1957	1970	1984	1984	1984	1991	1991	2000
United Kingdom	1993	ND	ND	ND	ND	2001	2001	1996

As can be seen from Table 5.1, the historical development of environmental policies in Europe has been very heterogeneous, with some countries leading the way (e.g. Denmark) and some following (e.g. the UK). It should be noted that several areas of EU environmental policy received a strong push as early as 1970s: energy taxes (France, Denmark, Germany, Finland, Norway with Sweden being the first in this area in 1957), and water pollution charges (with France and Sweden implementing the principles of environmental policy outlined in the theoretical literature). Sulphur taxes have seen a very wide discrepancy in the times of their introduction, in Norway they were implemented as early as 1970, however in France sulphur taxes appeared only in 1985, in Sweden in 1991, followed by Denmark (1996), Finland (1999), the Netherlands (2000), and the UK (2001). Although the economic mechanisms of sustainable waste management were the subject of a substantial debate in the 1970s, most EU countries did not introduce a landfill tax before the end of the 1980s, with Denmark and Austria showing leadership in this field. France joined the group in 1993, the Netherlands – in 1995, Finland and the UK – in 1996 and, surprisingly, Norway and Sweden – in 1999 and 2000 respectively. Carbon tax, which has been the focus of a lengthy discussion in the EU, is now present in a significant number of EU Member States; however France and Austria are not taking part due to lack of political acceptability. The world leader in this field was Finland (1990), followed by the Netherlands (1990), Norway and Sweden (both 1991), Denmark (1992), Germany (1999) and the UK (2001).

Figure 5.7 presents the spatial pattern of revenue generation through environmental taxes in major European countries expressed as share of GDP. The highest revenue from environmental taxes is currently received in Denmark, followed by the Netherlands, Cyprus and Bulgaria, which are, in turn, followed by the Scandinavian countries: Norway, Sweden and Finland, then Luxembourg, Hungary, Poland, Slovenia, Italy and Portugal.

In Fig. 5.8 we can see the dynamics of revenue due to environmental taxation in major EU countries from 1997 to 2008. It is clear that on the whole the tendency has been for environmental taxation revenue to increase, with some countries, like Netherlands, experiencing higher growth than others.

It remains largely to be seen how effective the introduction of environmental taxes has been for the reduction in the amounts of actual pollutants generated. This implies that much careful additional research is needed to establish the relative advantages of using environmental taxes for different types of pollutants and different countries: also their effect in conjunction with alternative measures. Two examples will illustrate the application of environmental taxes in this chapter. Figures 5.9 and 5.10 show the dynamics of the shares of the municipal solid waste (MSW) stream treated by the three major technologies: landfilling, incineration and recycling in Austria and the Netherlands. On the second axis the actual values of the landfill tax are depicted. Although the overall structure of consumption seems to be similar in Austria and the Netherlands and the amount of MSW is fairly similar stabilising at 600 kg per person per year, and both countries have seen considerable

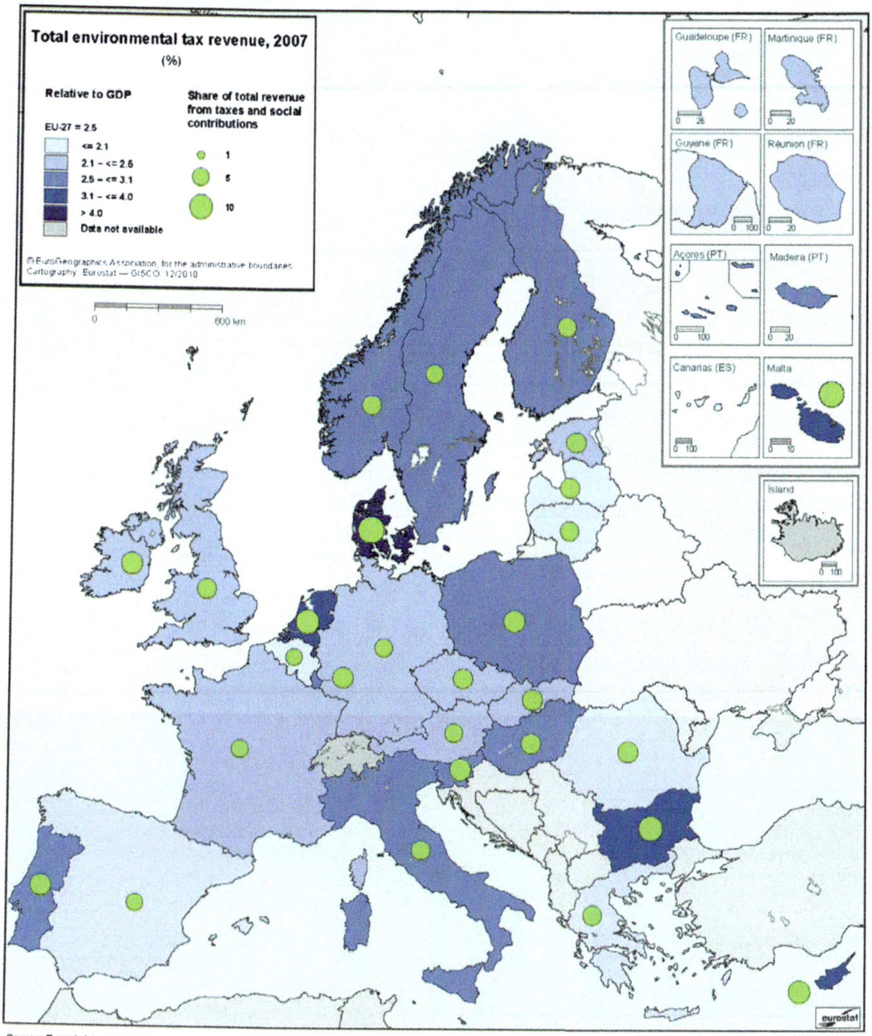

Source: Eurostat (env_ac_tax)

Fig. 5.7 Environmental tax revenues, EU

progress in terms of increasing the proportion of waste undergoing complex recycling, the Netherlands managed to decrease the proportion of waste which is landfilled even more than did Austria.

Perhaps one explanation of this difference lies in the fact that the rate of landfill tax in the Netherlands was raised to the level of over 60 Euros in 2000 whereas in Austria it was slightly less than 30 Euros at that time.

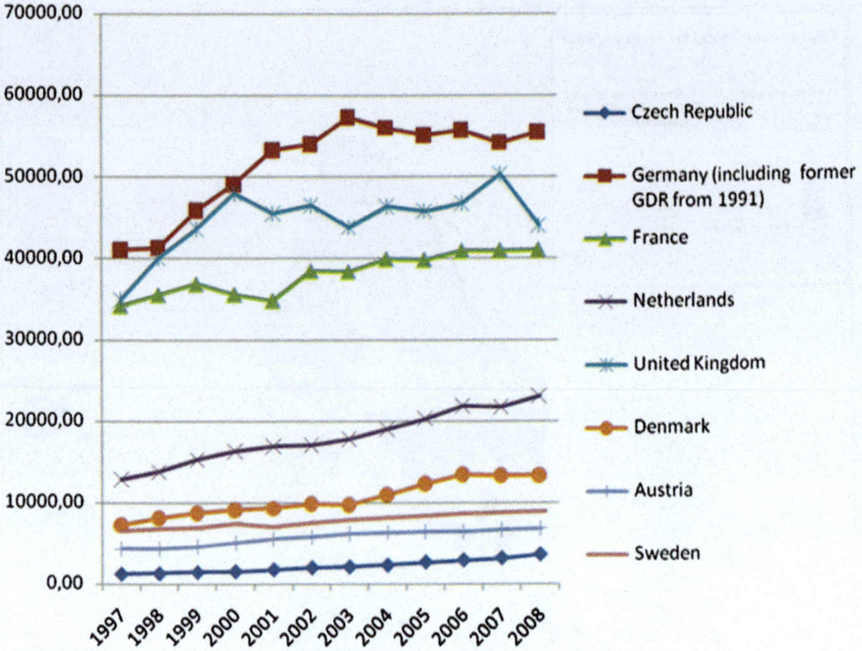

Fig. 5.8 Environmental taxation revenue, EU, 1997–2008 mln euro

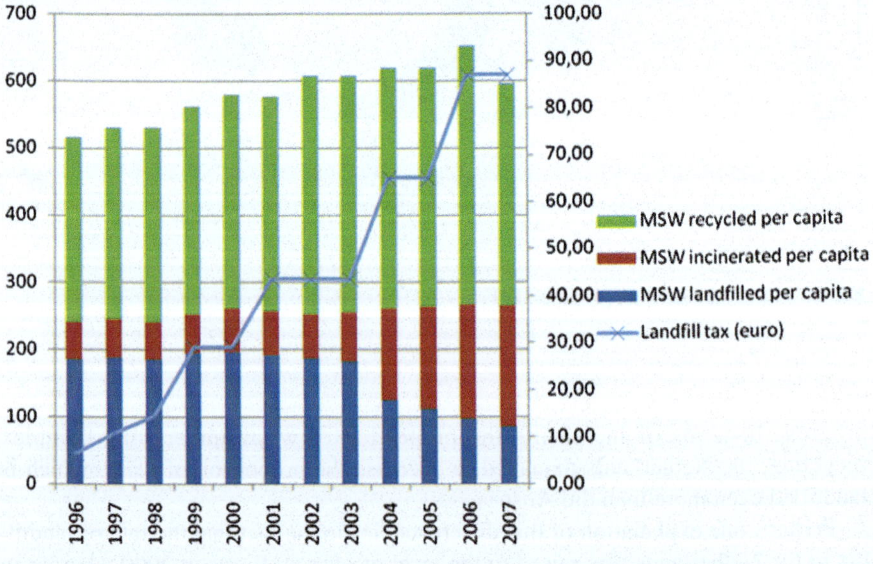

Fig. 5.9 Municipal solid waste treatment and landfill tax in Austria (1996–2008)

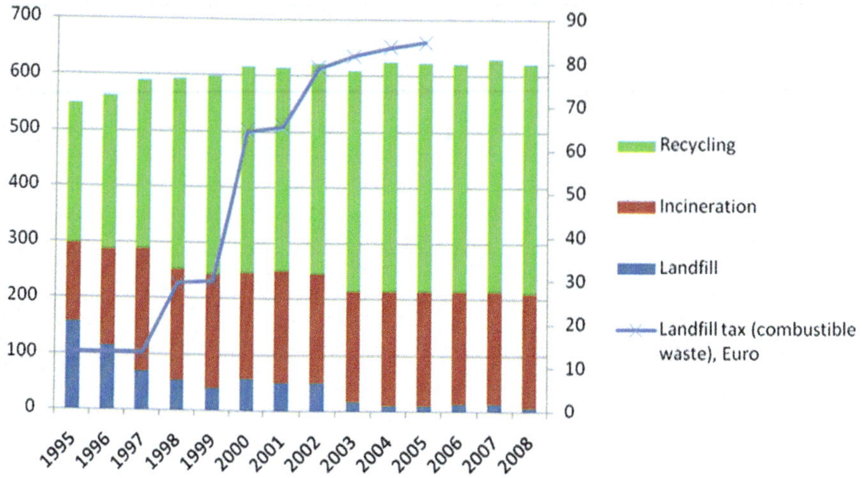

Fig. 5.10 Municipal solid waste treatment and landfill tax in the Netherlands (1995–2008)

References

Ayres RU and Kneese AV (1969) Production, Consumption, and Externalities *The American Economic Review* 59.3 282-297

Baumol WJ (1972) On Taxation and the Control of Externalities *The American Economic Review* 62.3 307-322

Baumol WJ and Oates WE (1971) The Use of Standards and Prices for Protection of the Environment *The Swedish Journal of Economics* 73.1 42-54

Coase RH (1960) The Problem of Social Cost *Journal of Law and Economics* 3 1-44

Hotelling H (1931) The Economics of Exhaustible Resources *The Journal of Political Economy* 39.2 137-175

Kneese AV (1971) Background for the Economic Analysis of Environmental Pollution *The Swedish Journal of Economics* 73.1 1-24

Leontief W (1970) Environmental Repercussions and the Economic Structure: An Input-Output Approach *The Review of Economics and Statistics* 52.3 262-271

Myrdal G (1973) Economics of an Improved Environment *World Development* 1.1-2 102-114

Pearce D (2002) An Intellectual History of Environmental Economics *Annual Review of Energy and the Environment* 27.1 57-81

Pigou A (1920) *The Economics of Welfare* London, Macmillan

Tietenberg TH (1973) Controlling Pollution by Price and Standard Systems: A General Equilibrium Analysis *The Swedish Journal of Economics* 75.2 193-203

Chapter 6
Economic Models and the Environment: Input–Output Analysis

Abstract This chapter explores the potential of combining two useful ecological-economic methods: input–output analysis and multi-criteria decision aid. By doing so, it assesses the sustainability of investment in various economic sectors, with the aim of minimising resource use and generation of emissions. The UK case is taken for the purpose of illustration, and (given the availability of the necessary data) this methodology might be applied in countries with various economic structures and specialisations. An environmentally extended static 123-sector UK input–output model is used, linking a range of physical flows (domestic extraction, use of water, and emissions of CO_2, CH_4, NO_x) with the economic structure of the UK. A range of environmentally adjusted forward and backward linkage coefficients has been developed, adjusted according to final demand, domestic extraction, publicly supplied and directly abstracted water, and emissions of CO_2 and NO_x. The data on the final demand adjusted and environmentally adjusted forward and backward linkage coefficients were used in a multi-criteria decision-aid assessment, employing a NAIADE method in three different sustainability settings. The assessment was constructed in such a way that each sector of the UK economy was assessed by means of a panel of sustainability criteria, maximising economic effects and minimising environmental effects. This type of multi-criteria analysis, could prove to be a valuable basis for similar studies, especially in the developing world, where trade-offs between economic development and environmental protection have been the subject of considerable debate.

Keywords Input–output analysis • Sustainability • Key sectors • MCDA • UK

Three Dimensions of Socio-Ecological Transformation

Three key elements seem to be crucial for the socio-ecological transformation if our society is to reach sustainable development and overcome growing energy and resource requirements and rising volumes of emissions and wastes, to facilitate

S.E. Shmelev, *Ecological Economics: Sustainability in Practice*,
DOI 10.1007/978-94-007-1972-9_6,

change to renewable energy sources and conservation of biodiversity. Firstly, this is a the concept of industrial ecology (Graedel and Allenby 2002), which highlights the importance of intersectoral flows of matter and energy required for the production of goods and services, analysed in detail throughout the whole lifecycle of the given product, service or a regional or national system. Secondly, it is the system of tools for decision making (Söderbaum 2000) based on multicriteria methods which, applied at different levels, would shift the patterns of decision making towards more socially equitable and more environmentally friendly as well as more economically sound decisions. Thirdly, it is a system of macroeconomic goals or sustainability assessment methods which dominate on the macroeconomic scene. For a very long time GDP has been the key variable, which was at the heart of macroeconomic policies all over the world. Due to the efforts of ecological economists, and, especially, Herman Daly (2000) a new vision was proposed, the vision of sustainable development as a qualitative creative change, as opposed to quantitative growth. The idea of incommensurability of values, incorporated in the concept of sustainable development, have lead to the development of new alternative sustainable development assessment approaches (Shmelev and Rodríguez-Labajos 2009).

Environmentally Extended Input-Output Analysis

Different countries started to develop input–output tables after the publication of the first balance of the national economy of the USSR and its subsequent criticism by Leontief (Table 6.1). Tables for the USA (1919, 1929, and 1947) followed. Later Norway (1948), the Netherlands (1948), Japan (1951) and the UK (1954) joined the process. With a little delay, Hungary (1957), Poland (1957), USSR (1959) and Brazil (1959) continued the trend. The resolution of the input–output tables varied significantly: if the first tables for the USA contained 44 and 41 sectors respectively, the Netherlands – 35 sectors, it was soon realised that increasing the amount of detail allows an unprecedented capacity to understand and manage the complexity of intersectoral linkages. Subsequently, tables for the USA included 400 sectors, Japan – 399 sectors; Estonia – 239 sectors; Lithuania – 239 sectors; Belorussia (500 sectors).

The first tables to appear in the USSR after the WWII, including the tables for Estonia, Latvia and Lithuania (239 sectors, 1961) have been described in Jasny (1962) and Kossov (1964). The first Dutch input–output tables to appear have been reviewed by Rey and Tilanus (1963), the first international comparative analysis of the economies of the USA, Japan, Norway, Italy, Spain using input–output tables was offered by Simpson and Tsukui (1965).

Environmentally extended input–output applications started to develop in the 1970s (Table 6.2) following the original publication by Leontief and they covered the following issues: energy and the environment (Carter 1974, 1976, Gay and Proops 1993, Herendeen and Tanaka 1976, Park 1982, Polenske and Lin 1993,

Table 6.1 Input-output tables published in world countries

Country	Year, referring to	Number of sectors
USSR	1923/1924	12 sectors
USA	1919	44 sectors
USA	1929	41 sector
USA	1947, 1958, 1963	400 sectors, 480 intermediate sectors
Norway	1948	175 sectors
Netherlands	1948–1957	35 sectors
Japan	1951, 1973, 1976	399 intermediate sectors (2005)
UK	1954, 1961	123 intermediate sectors
Hungary	1957	40 sectors
Poland	1957	20 sectors
USSR	1959	83 sectors
Brazil	1959	32 sectors
Brazil	1969, 1970	87 sectors
Estonia	1961	239 sectors
Lithuania	1961	239 sectors
Canada	1961	250 industries
Belorussia	1962	500 sectors
China	1973	61 sector,
China	1997	124 commodities
Australia	1974	135 sectors
OECD	1972, 1977, 1982	48 sectors

Proops 1977, 1984); materials balance and materials flows (Duchin 2004, Giljum 2004, Hoekstra 2005, Suh 2009, Tukker et al. 2009); water (Anderson and Manning 1983, Dietzenbacher and Velázquez 2007, Lenzen 2009, Lenzen and Foran 2001, Wang and Wang 2009, Wang et al. 2005); waste (Duchin 1990, 1994, Kondo and Nakamura 2005, Leontief 1977a, Nakamura 1999, Nakamura and Kondo 2002, 2006) and the environmental policy analysis (Gutmanis 1975). The UN global model project has significantly stimulated interest in the analysis of the environmental consequences of economic development and effects of technological innovation (Ayres and Shapanka 1976, Carter and Petri 1979, Leontief 1977b, Leontief and Duchin 1986, Petri 1977). Substantial projects focused on the application of input–output analysis to national economies for policy analysis have been started in various countries including the UK (Barker 1981, Barker et al. 1980, Stone 1984). Dynamic input–output analysis has become one of the most interesting subjects for economic research (Duchin and Szyld 1985, Raa 1986, Vogt et al. 1975). Environmentally extended input–output analysis of the changes in the world economy has been carried out by (Duchin 1986, Fontela 1989, Leontief and Duchin 1986, Schäfer and Stahmer 1989). Later, this framework was extended to include material flows (Duchin 2004), other pollutants (Duchin 1994, 1998) and different types of waste (Nakamura 1999). The most recent applications of extended input–output analysis today include an environmental key sector analysis by Manfred Lenzen (Lenzen 2003), and econometric extended-input–output models of the UK and the European Union (Barker et al. 2007a, b).

Table 6.2 Major contributions in environmentally extended input–output analysis

Author, year	Country of application	Sectoral dimensions	Extensions
(Leontief 1970)	N/A	2×2	1 pollutant, agriculture and manufacturing
(Leontief and Ford 1972)	USA	90 sectors	5 residuals, 1 recipient (air), 11 final demand categories,
(Leontief 1974)	World		45 sectors, 40 minerals and fuels, 30 pollutants
(Forsund and Strom 1976)	Norway	86 sectors	35 types of residuals, 28 final demand categories
(Proops 1977)	UK	3×3	Energy intensities
(Barker 1981)	UK	40 sectors	Econometrics, annual time series 1954–1979, and cross-section data in the form of input–output tables 1954, 1963, 1968, 1974.
(Luptáčik and Böhm 1994)	N/A		MCDA, trade-off between economic goals and the quality of the environment
(Kananen et al. 1990)	Finland	17 sectors	MCDA, emergency management
(Duchin 1992)	N/A	4×4	Industrial ecology
(Gay and Proops 1993)	UK	38 sectors	CO_2
(Sonis and Hewings 1998)	Indonesia	5 sectors	Structural path analysis, SAM
(Nakamura 1999)	Netherlands	20 sectors	Waste, recycling and CO_2 emissions
(Ferrer and Ayres 2000)	France	30	Waste, remanufacturing
(Moffatt and Hanley 2001)	Scotland	28 sectors	12 pollution types
(Hoekstra and van den Bergh 2002)	N/A	N/A	MFA and structural decomposition analysis
(Aroche-Reyes 2003)	Mexico	27 sectors	Qualitative analysis of economic structures
(Lenzen 2003)	Australia	134 sectors	Environmentally adjusted linkage coefficients
(Giljum and Hubacek 2004)	Germany	3×3	Primary material inputs
(Lantner and Carluer 2004)	France	36×36	Spatial dominance: 6 regions, 6 sectors each
(Suh 2005b)	N/A		MFA and energy
(Suh 2005a)	USA	500 sectors	Life cycle input–output
(Peters and Hertwich 2006)	Norway	49 sectors	Internatonal trade, embodied CO_2
(Cardenete and Sancho 2006)	Spain, 1995	10 sectors	SAM
(Moran and del Rio Gonzalez 2007)	Spain	44 sectors	CO_2 emissions

Figure 6.1 contains a schematic description of material and energy flows in the national economy. The outer, light green box depicts the boundaries of the environment system, with a yellow box "Energy", responsible for the transfer of solar energy to ecosystems and humans. The inner, dark yellow box represents the economic system, forming part of a wider environmental system, and constrained by the limitations of the environmental system. The principle of embeddedness of the

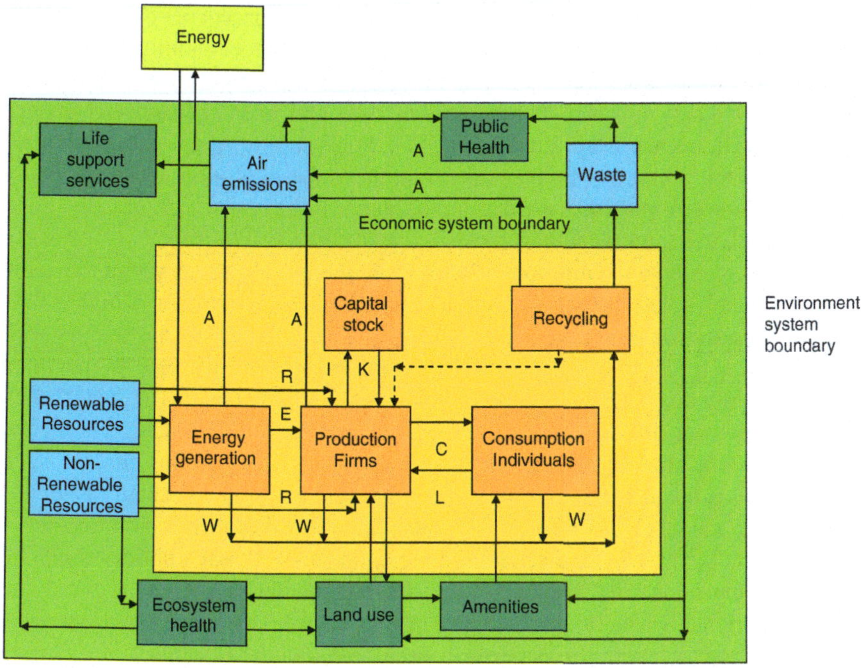

Fig. 6.1 Economy-environment interdependence (Modified and enhanced from Common and Stagl 2005)

economic system in the environmental system became the subject of considerable debate and a lot of attention from such pioneers of ecological economics as Herman Daly (2000). The dark ochre boxes represent fundamental economic activities, such as energy generation, production, consumption, and accumulation of capital stock and recycling, a new type of economic activity, designed to bring economic systems closer to the sustainable path and to emulate natural ecological metabolic processes. Light blue boxes in the chart represent the stocks of renewable and non-renewable resources taken from the natural environment, and emissions and waste emitted to the environment as a result of the functioning of the economic system. Emissions to water and some other factors are not considered here for the sake of simplicity. The dark green boxes situated outside the economic system represent the key factors which should be taken in the account, when analysing the future development of the economy: life support services, ecosystem services, public health, visual and other amenities, and land use generally. It is a very rough classification of the types of impacts which might be adjusted in each individual case. It was successfully applied to the analysis of the sustainability of regional waste management systems (Shmelev and Powell 2006). When such a range of aspects of the development of a given regional or national system is considered, it seems desirable to use special multicriteria methods to support decisions at all levels of the decision making process, which will be covered in the next section of the paper.

In his pioneering article Lenzen (2003) introduced the concept of environmentally important paths, linkages and key sectors in the macroeconomic framework. Historically, Rasmussen was the first to introduce the concept of forward and backward inter-industry linkages as measures of structural interdependence (Hewings et al. 1989, Hirschman 1958, Rasmussen 1956, Sonis and Hewings 1999, Sonis et al. 1995). Lenzen (2003) for the first time introduced the idea of an environmentally adjusted forward and backward inter-industry linkages, which are designed to highlight the sectors, which have higher than average propensity to cause resource extraction and emissions across the economy. The sectors with the value of the forward linkage coefficient higher than one tend to produce a higher than average impact "downstream" in their supply chain. Similarly, the sectors with the backward linkage coefficient larger than one tend to produce higher than average impact on the economy "upstream" in their supply chain. The sectors with the value of both forward linkage coefficient and backward linkage coefficient higher than one are usually referred to as the "key sectors". In this chapter, such an approach is taken one step further and applied to the environmentally extended input–output model of the UK economy, comprising 123 sectors and additional flows of domestically extracted materials, directly abstracted and publicly supplied water and emissions of CO_2, NH_4, NO_x. Environmentally adjusted forward and backward oriented linkages are calculated here for all the 6 mentioned environmental aggregates and illustrate the pattern of direct and indirect effects of investing in particular sectors of the UK economy as of 2000.

The particular innovative aspect of the analysis in this chapter is the subsequent treatment of the derived forward and backward linkage coefficients with the help of multicriteria decision aid (MCDA) tools, which helps to identify the most "sustainable" sectors of the British economy in terms of their power to stimulate economic development, producing at the same time, minimal environmental effects across the national economy.

Integration of economic input–output analysis and information on the physical flows passing through the economy allow us to undertake a detailed analysis of the structural physical links in the economy with the aid of environmental key sector analysis. Taking into account physical flows is a major advantage of this approach, as it allows us to look beyond the simple monetary value of transactions in the input–output table and explore the rich complexity of physical linkages which exist in the economy. This will prove extremely beneficial in analysing the economy-wide environmental effects of government investment programmes in times of crisis.

Modelling the UK Economy

The static UK input–output model created by the author was used in this paper with extensions of resource and environmental flows. The input–output 123 sector tables referring to the year 2002 were obtained from the UK Office for National

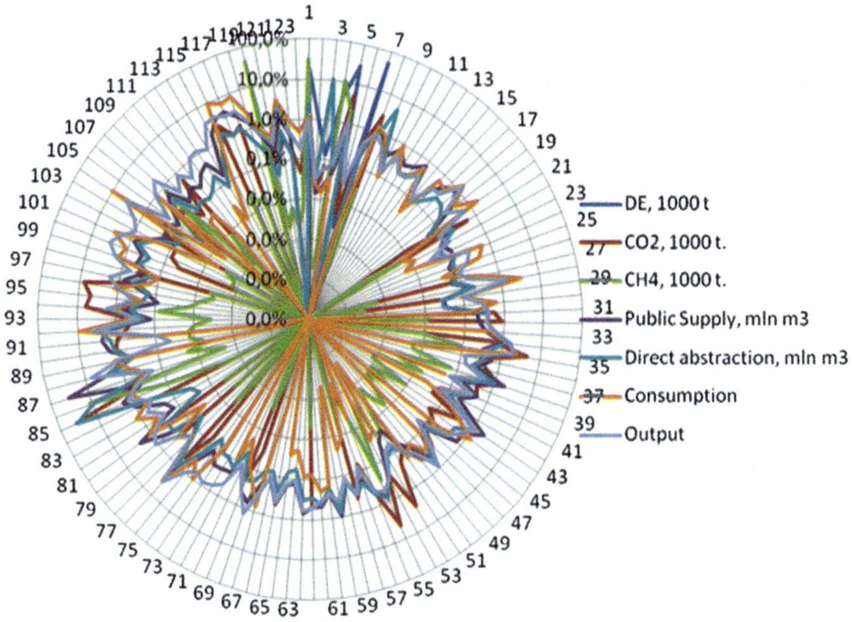

Fig. 6.2 Economic and physical flows in the UK economy (123 sectors), 2002

Statistics, the full sector classification can be seen in the Annex 2 of this paper. It should be noted that the results of the subsequent analysis should be treated as a first approximation, because not all elements of the UK input–output table are available to the public due to confidentiality regulations. The water accounts of the UK had to be adjusted for they do not provide the necessary detail and further disaggregation was carried out by the author. The data on material flows has been obtained from the MOSUS project, where the author took an active part by developing the global database of material flows for 1980–2003, which included all countries of the world and around 400 types of flows according to EU guidelines (Shmelev and Giljum 2004). Data on UK CO_2 emissions as well as data on CH_4 and NO_x emissions come from the UK Office for National Statistics.

An integrated illustration of economic and environmental flows in the UK economy is depicted in Fig. 6.2. Each economic sector (the names and respective numbers can be found in Annex 2) is characterised by the share of its domestic extraction of natural resources, publicly supplied and directly abstracted water, emissions of CO_2, CH_4, consumption and economic output, presented on the logarithmic scale. Table 6.3 presents the most relevant sectors (with shares greater than 5%) in terms of their direct environmental and economic effects, with respective percentages of the total flow.

Table 6.3 Direct environmental and economic sectoral impacts

Dimension	Sectors	Share
Domestic extraction		
	Other mining and quarrying	49.6%
	Oil and gas extraction	28.0%
	Agriculture	17.2%
Water publicly supplied		
	Water supply	32.4%
Water directly abstracted		
	Electricity production and distribution	33.0%
	Fishing	10.8%
	Gas distribution	9.0%
	Fish and fruit processing	5.1%
CO_2		
	Electricity production and distribution	36.0%
	Air transport	7.6%
	Other land transport	6.0%
CH_4		
	Sewage and sanitary services	42.5%
	Agriculture	31.5%
	Gas distribution	11.3%
	Coal extraction	10.9%
Consumption		
	Letting of dwellings	9.9%
	Public administration and defence	9.8%
	Hotels, catering, pubs, etc.	8.8%
	Health and veterinary services	8.1%
Output		
	Construction	6.7%

Environmentally Adjusted Forward and Backward Linkages in the UK Economy

Figures 6.3 and 6.4 depict final demand and CO_2 adjusted coefficients of forward and backward linkages, which characterise the national economy of the United Kingdom in 2002 from the point of view of economic and environmental intensities of the physical links among different sectors. In Fig. 6.3 all sectors are grouped into four clusters: key sectors, backward linkage oriented, forward linkage oriented, and weak oriented sectors. For key sectors the respected value of both forward and backward linkage coefficient is greater than 1. The corresponding sector names and numbers can be found in Annex 2.

We can see from Fig. 6.3, that in the pure economic sense, which corresponds to traditional economic thinking historically applied in different countries, the sectors associated with the strongest economic links with the rest of the economy, capable

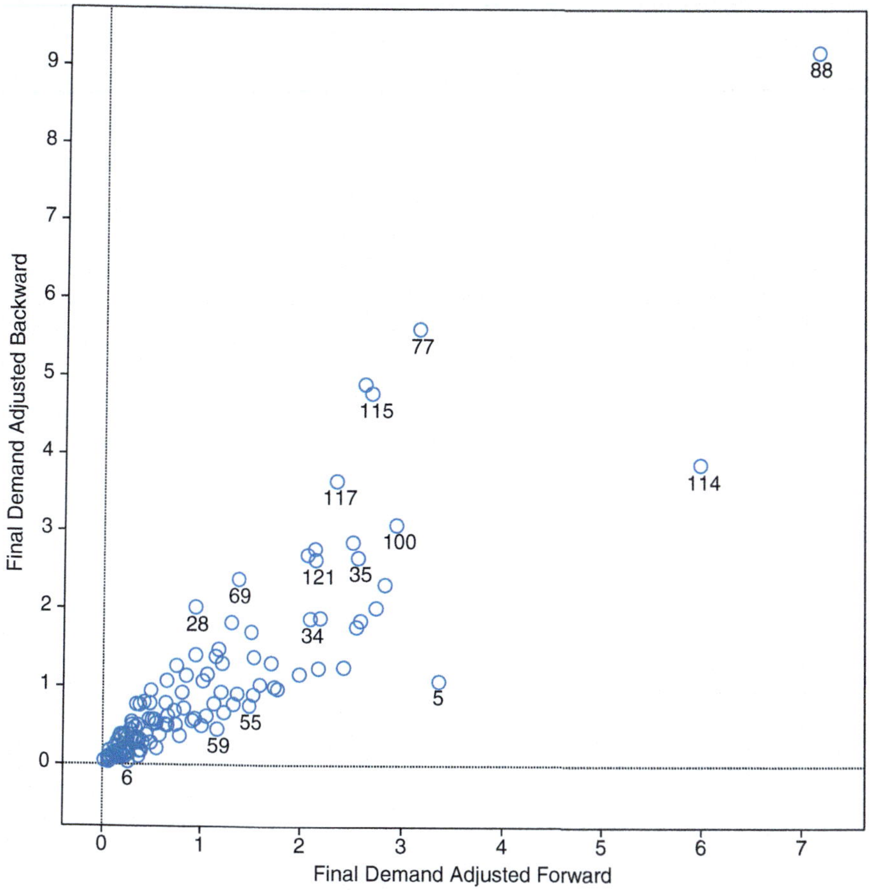

Fig. 6.3 Final demand adjusted forward and backward linkage coefficients, labelled by sector UK, 2002

of stimulating economic development, in the UK in 2002 were construction, other business services, motor vehicles, hotels and catering, public administration and defence, health and veterinary services, banking and finance.

CO_2 adjusted forward and backward linkage coefficients for the major industries depicted in Fig. 6.4, give us a different picture. The most CO_2 forward and backward linked sector is Electricity production and distribution, other key sectors in relation to CO_2 impacts in the UK economy are Construction, Coke ovens, Refined petroleum and nuclear fuel, Motor vehicles, Iron and Steel, Air Transport, Oil and Gas Extraction and several others. It is quite natural, that the forward linkage coefficient for Oil and Gas Extraction is much higher than the backward linkage due to the role, that oil and gas play as fuels in the transport and other sectors. The reverse applies to air transport, due to the amount of fuel that is used on flights.

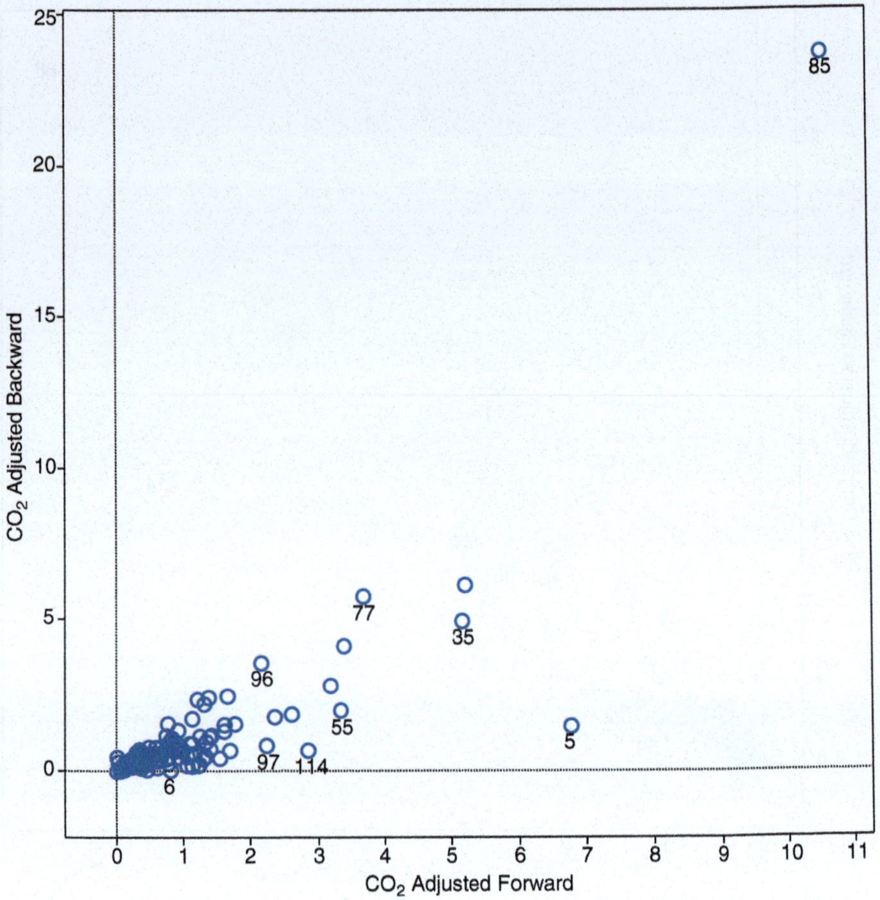

Fig. 6.4 CO_2 adjusted forward and backward linkage coefficients, labelled by sector, UK, 2002

Key sectors in the environmental sense, when domestic extraction is taken as a basis for weighting the coefficients (Fig. 6.5), were the following: Other mining and quarrying, Construction, Coke ovens, refined petroleum and nuclear fuel, Oil and Gas Extraction, Agriculture, Electricity production and distribution and some others. For these sectors, additional economic activity would mean higher than proportional resource extraction impacts further up and down the supply chain; the respective coefficients are shown on the chart's axis. For example, for the Oil and Gas Sector, domestic extraction, the adjusted forward linkage coefficient is 9.53 and backward linkage coefficient is 5.16. This means that oil and gas extraction generates forward oriented extraction impacts that are 9.53 times higher than the domestic extraction impact of oil and gas alone. Respective interpretation can be applied to the backward linkage coefficients.

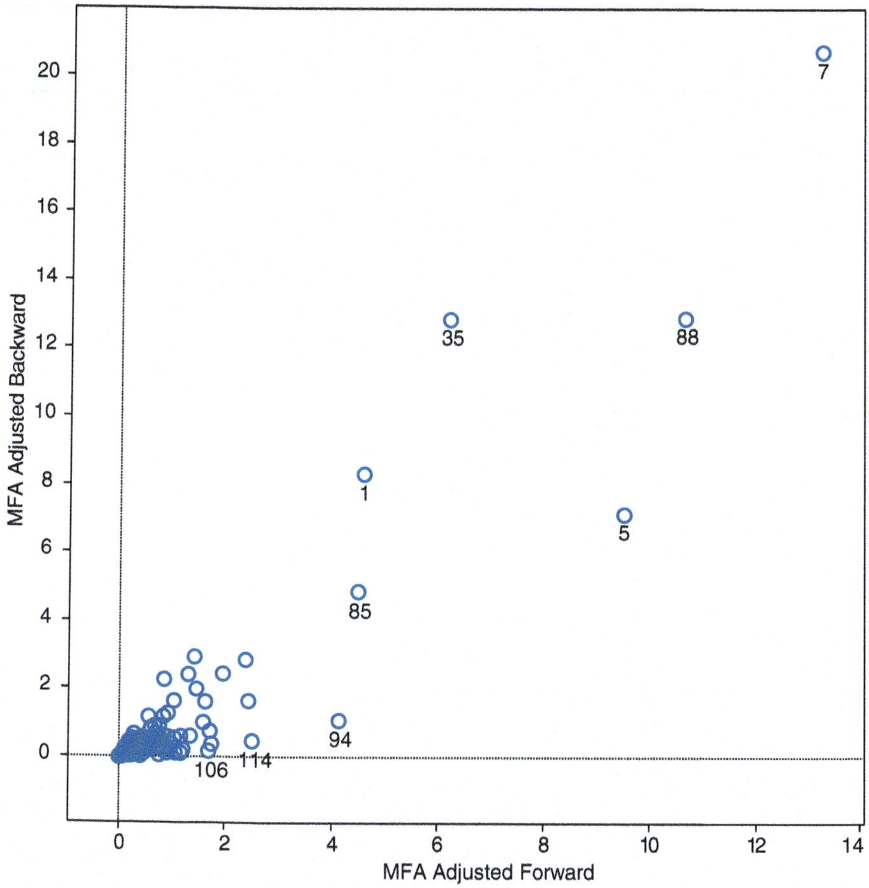

Fig. 6.5 DE adjusted forward and backward linkage coefficients, labelled by sector, UK, 2002

When the economic system is considered from the point of view of associated emissions of NO_x (Fig. 6.6), the following pattern is produced. The sector, characterised by the largest potential to influence the generation of NO_x emissions in the UK in 2002 was Water Transport, followed by Computer Services, Electricity Production and Distribution, Construction, Motor Vehicles, Non-Ferrous Metals, Coke Ovens etc., Other Land Transport and some others.

When the economic system is considered from the point of view of associated water flows (directly abstracted and publicly supplied) the following pattern emerges. In the case of publicly supplied water the strongest key sectors are: Water Supply, Motor Vehicles, Organic Chemicals, and Construction etc. For directly abstracted water the "key sectors" are: Electricity Production and Distribution, Fish and Fruit Processing, Fishing and so on (Figs. 6.7 and 6.8).

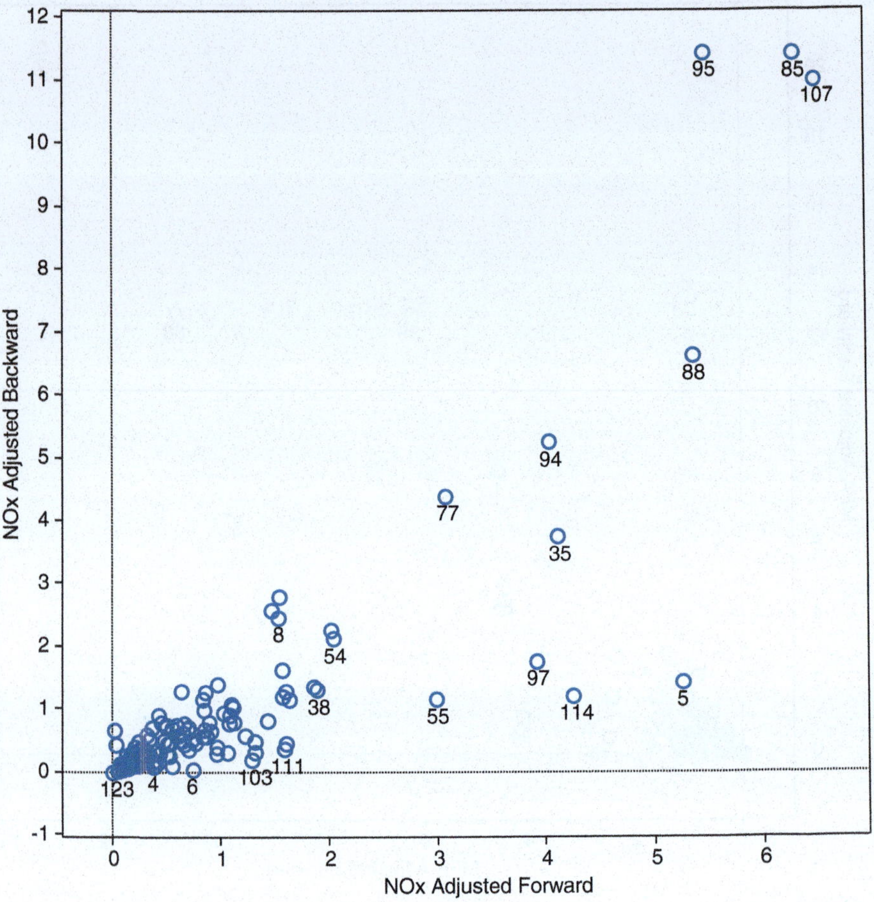

Fig. 6.6 NO_x adjusted forward and backward linkage coefficients, UK, 2002

Macro Sustainability Assessment with MCDA

There is a wide spectrum of aspects which should be taken into account when discussing sustainability: the UN system of indicators of sustainability comprises 96 indicators with a core of 50 indicators divided into 14 themes: Poverty, Governance, Health, Education and Demographics, Natural Hazards, Atmosphere, Land, Oceans, Seas and Coasts, Freshwater, Biodiversity, Economic Development, Global Economic Partnership, and Consumption and Production Patterns. Therefore a whole new class of methods is required to address sustainability problems at the local, regional and national level, taking a range of criteria into account simultaneously. Such methods are usually referred to as multicriteria decision aid (MCDA) methods and have been developed within many different schools: in France, in the

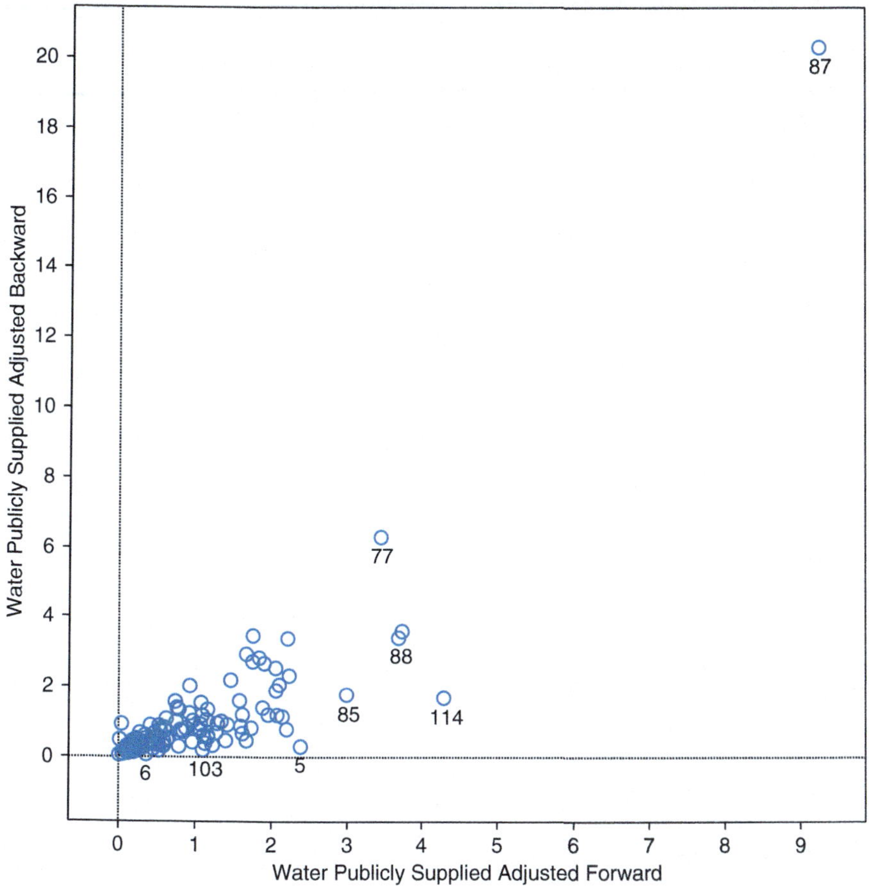

Fig. 6.7 Publicly supplied water adjusted public forward and backward linkage coefficients, UK, 2002

Netherlands, in the USA, in Russia and several other countries. Methodological work in this field has been done by (Ferrer and Ayres 2000) applying these methods to regional problems, (Roy 1985), the author of one of the most famous families of multicriteria methods, outranking methods "ELECTRE"; (Janssen 1993), who developed a decision support tool called DEFINITE, the author of the method, called NAIADE, based on fuzzy logic. There is an extensive body of work covering the use of multicriteria methods in decision making. A range of multicriteria programming methods has been developed to deal with well structured and quantitatively described problems. Numerous applications of MCE exist for regional problems, e.g. waste management (Shmelev and Powell 2006) or renewable energy (Madlener and Stagl 2005). The novel application of such methods to macro-sustainability assessment has been offered in (Shmelev and Rodríguez-Labajos 2009).

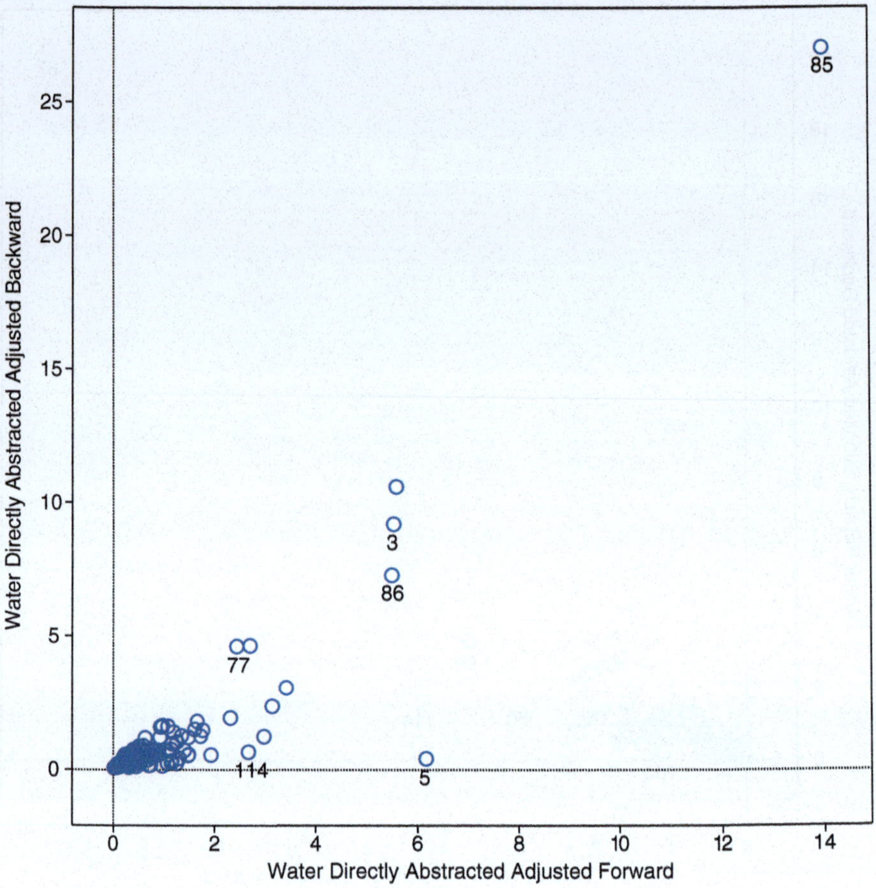

Fig. 6.8 Directly abstracted water adjusted forward and backward linkage coefficients, labelled by sector, UK, 2002

The perspective of the MCDA presents a new paradigm, which is different from the classical goal of finding an optimal solution subject to a set of constraints characteristic of operations research.

A Novel Approach to Imprecise Assessment and Decision Environment (NAIADE) is a discrete multicriteria method whose impact (or evaluation) matrix may include either crisp, stochastic or fuzzy measurements of the performance of alternative and with respect to a judgement criterion (Munda 1995, 2005). No traditional weighting of criteria is used in this method. The whole procedure can be divided in three main steps:

- pairwise comparison of alternatives;
- aggregation of all criteria;
- evaluation of alternatives.

The method is based on the concept of fuzzy preference relation. If A is assumed to be a finite set of N alternatives, a fuzzy preference relation is an element of the N×N matrix $R=(r_{ij})$, i.e.:

$$r_{ij}=\mu R(a_i, a_j) \text{ with } i,j=1,2,\ldots,N \text{ and } 0<=r_{ij}<=1$$

$r_{ij}=1$ indicates the maximum degree of preference of a_i over a_j; each value of r_{ij} in the open interval (0.5, 1) indicates a definite preference of a_i to a_j (a higher value means stronger intensity); $r_{ij}=0.5$ indicates the indifference between a_i and a_j.

Six different fuzzy relations are simultaneously considered:

1. much greater than (>>)
2. greater than (>)
3. approximately equal to (~)
4. exactly equal to (=)
5. less than (<)
6. much less than (<<).

Given the information on the pairwise performance of the alternatives according to each single criterion, these evaluations are aggregated in order to take all criteria into account simultaneously. As a final result, the method creates the webs of domination relationships among alternatives, and presents them in a useful graphical form. The main distinct feature of NAIADE, which was particularly important for our analysis was the capacity to change the degree of sustainability (parameter α), from weak to strong to simulate the changes in perspective on the degree of compensation allowed among the criteria. For a more detailed description of the NAIADE method the interested reader is referred to (Munda 1995).

Application of MCDA Methods for Sustainability Analysis

The only known application of the MCDA tools in the input–output context belongs to (Luptáčik and Böhm 1994). The authors use the input–output model as a basis for a multicriteria optimization programme to identify the optimal structure of output, which minimises environmental effects under the constraints of primary input. Our approach is different in that, working with the real input–output model of the UK economy, we use environmentally adjusted forward and backward linkage coefficients to find the most environmentally sustainable and economically viable industries. The data obtained as a result of the calculation of forward and backward linkage coefficients has been used in the multicriteria decision aid (MCDA) system NAIADE, which is an outranking MCDA tool, capable of handling various types of data including interval, crisp, stochastic and fuzzy elements. The method produces webs of domination relationships, for weak, neutral and strong sustainability settings, which can be seen in detail in Annex 1.

Each of the UK economic sectors was taken into account in the MCDA assessment with eight coefficients each respectively: forward and backward linkage coefficients adjusted for final demand, domestic extraction, CO_2 and NO_x. It was assumed that it

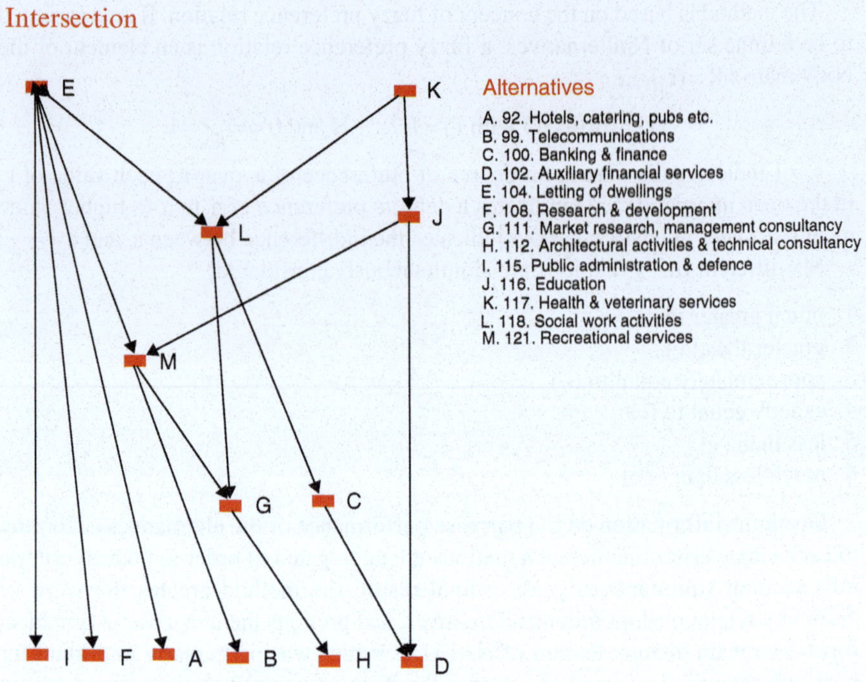

Fig. 6.9 The web of domination relationships, UK most sustainable economic sectors (final demand, domestic extraction, CO_2, NO_x, adjusted linkages), 2002

is in the interests of society to maximise the final demand adjusted economic multiplier characteristics of certain sectors, at the same time minimising direct and induced material extraction, CO_2 emissions and NO_x emissions. The web, depicted in Fig. 6.9 could be interpreted in the following way: economic sectors E (104, Letting of Dwellings) and K (117, Health and Veterinary Services) turned out to be the most sustainable, dominating all the other sectors and occupying the top positions in the web under the neutrality assumption about sustainability (the results change in weaker or stronger sustainability settings) .The sectors L (118, Social work activities) and J (116, Education) were the next from the top in terms of sustainability criteria outlined here, while the sector M (121, Recreational Services) occupied the next "layer in the hierarchy", however L and J or L and M are not comparable between themselves (there are no connecting arrows between the two in any pair). The existence of an arrow in such a diagram shows the existence, and the direction of the arrow — direction of the domination relationship that exists between the alternatives in question, the lack of such an arrow points to incomparability.

The sectors G (111, Market research, management consultancy) and C (100, Banking and finance) occupied the layer at the bottom of the middle of the web, and finally, the sectors I (115, Public administration and defence), F (108, Research

Table 6.4 Top sustainable sectors in the UK economy under different assumptions, 2002

Scenario	Top 10 sectors
α = 0.1 (weak sustainability)	104 Letting of dwellings
	121 Recreational services
	118 Social work activities
	116 Education
	102 Auxilary financial services
α = 0.5 (neutrality)	104 Letting of dwellings
	117 Health and veterinary services
	116 Education
	121 Recreational services
	118 Social work activities
α = 0.9 (strong sustainability)	115 Public administration and defence
	92 Hotels, catering, pubs, etc.
	117 Health and veterinary services
	104 Letting of dwellings
	118 Social work activities

and Development), A (92, Hotels, catering and pubs), B (99, Telecommunications), H (112, Architectural activities etc.), and D (102, auxiliary financial services) occupied the bottom of the web of domination relationships and were the least sustainable of the set of sectors identified in this example. It should be stressed that the parameter α plays a crucial role in determining the shape of the resulting web of domination relationship.

Table 6.4 presents the summary of the results in terms of the top sustainable sectors in all the settings.

For a discussion of the differences between strong and weak sustainability in the NAIDE applications, the reader is referred to (Shmelev and Rodríguez-Labajos 2009), the key difference being the ease of compensation among the sustainability criteria in the case of weak, and the strong complementarity and lesser compensation in the strong sustainability setting.

It can be seen from Table 6.4 that such sectors as 116 (Education), 117 (Health and Veterinary Services), 118 (Social Work Activities), 104 (Letting of Dwellings), 121 (Recreational Services) feature prominently in almost all sustainability settings, and are those sectors that truly provide the basis for the sustainable development of the United Kingdom both in the sense of direct effects and indirect effects, thereby not inflicting heavy resource use or pollution load across the whole spectrum of economic sectors. This result is extremely important for the preparation of economic recovery programmes by the UK Government, focused in the neo-Keynesian sense on stimulating economic recovery. One would hope that this economic crisis might be seen as an opportunity not only to concentrate on pure economic recovery, but also on wider resource use and environmental impacts and on the strategic environmental modernisation of the economy. In any case, any reduction in educational or health care budgets according to these results is completely unjustified and would be harmful for the economy in the long run, especially if from a sustainability perspective.

Conclusions

As our application shows, the combination of various approaches proves to be especially fruitful. In our case, environmentally extended input–output analysis has been combined with multi-criteria decision aid to identify the sectors, that are "most sustainable" both in terms of direct and indirect impacts. The unique aspect of this application lies in its use of environmentally adjusted forward and backward linkage coefficients which show the effects produced through the web of intersectoral linkages. This chapter presents a novel way of assessing the relative sustainability of investment in particular economic sectors from the point of view of resource use and generation of emissions. The research carried out can be disaggregated into the following three steps: an environmentally extended static 123 sector UK input–output model has been created, which linked a range of physical flows: domestic extraction, use of water, emissions of CO_2, CH_4, NO_x, with an economic structure of the UK. Secondly, following a range of environmentally adjusted forward and backward linkage coefficients has been developed, with a particular focus on final demand, domestic extraction, CO_2 emissions and NO_x emissions adjusted coefficients. Then, the data on the final demand and environmentally adjusted forward and backward linkage coefficients is used in a multicriteria decision aid (MCDA) assessment, employing a Novel Approach to Imprecise Assessment and Decision Environments (NAIADE) method in three different sustainability settings: weak sustainability, strong sustainability and a neutral setting. The assessment is set in such a way that each of the 123 sectors of the UK economy is compared with the others using a panel of sustainability criteria, with final demand adjusted coefficients aimed at their maximum and, environmentally adjusted – at their minimum values.

The results show that the following sectors:

117 Health and Veterinary Services
104 Letting of Dwellings
116 Education
121 Recreational Services
101 Insurance and Pension Funds
118 Social Work Activities
99 Telecommunications

appear with relative stability within the top 10 most sustainable sectors of the UK economy from the point of view of both direct and indirect effects in the strong sustainability, weak sustainability and neutral assessment.

It can be seen that the analysis conducted might be a justification for a substantial governmental investment programme which could not only stimulate the development of the economy, but also reduce the direct and indirect environmental consequences of such development. Such a programme seems to be particularly desirable in the conditions of the current economic crisis, which in our opinion presents a challenge and at the same time offers an opportunity for the reorientation of governmental investment priorities towards more sustainable industries. Unfortunately this particular aspect of the problem is not currently being discussed by any of the political parties in the UK (Figs. 6.10–6.12).

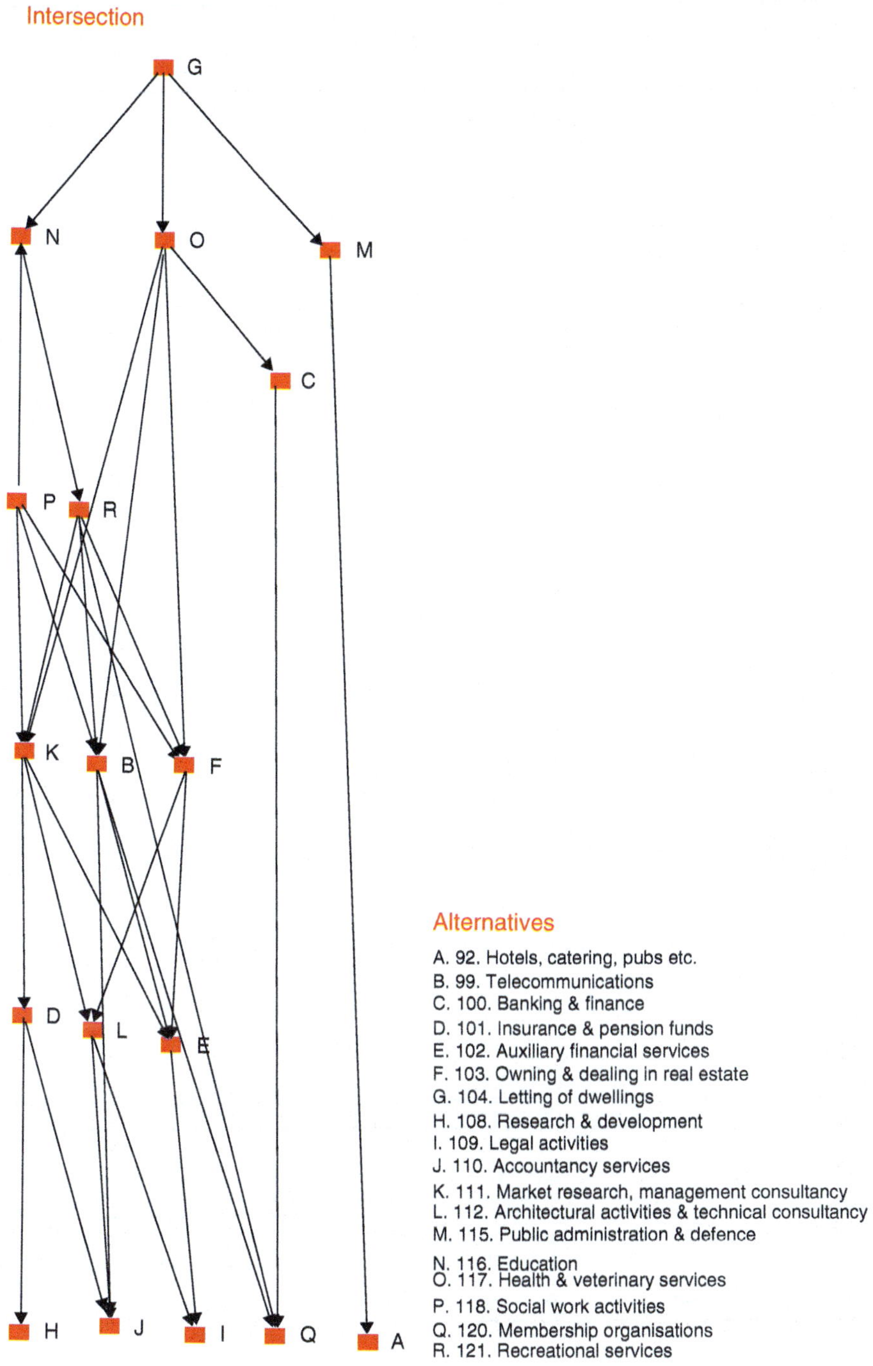

Fig. 6.10 The most sustainable sectors, UK, 2002, $\alpha = 0.1$ – weak sustainability setting

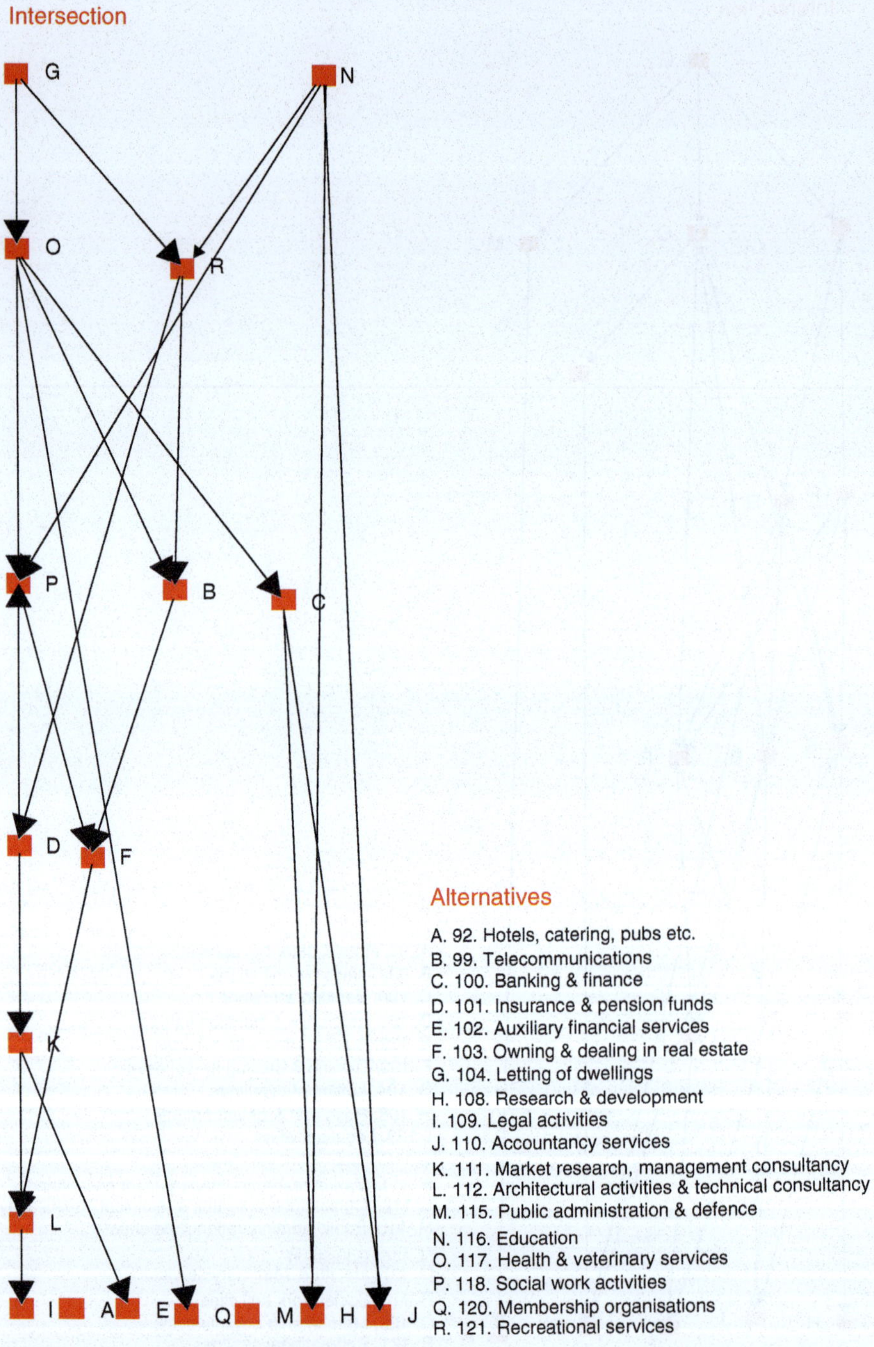

Fig. 6.11 The most sustainable sectors, UK, 2002, $\alpha = 0.5$ – neutrality setting

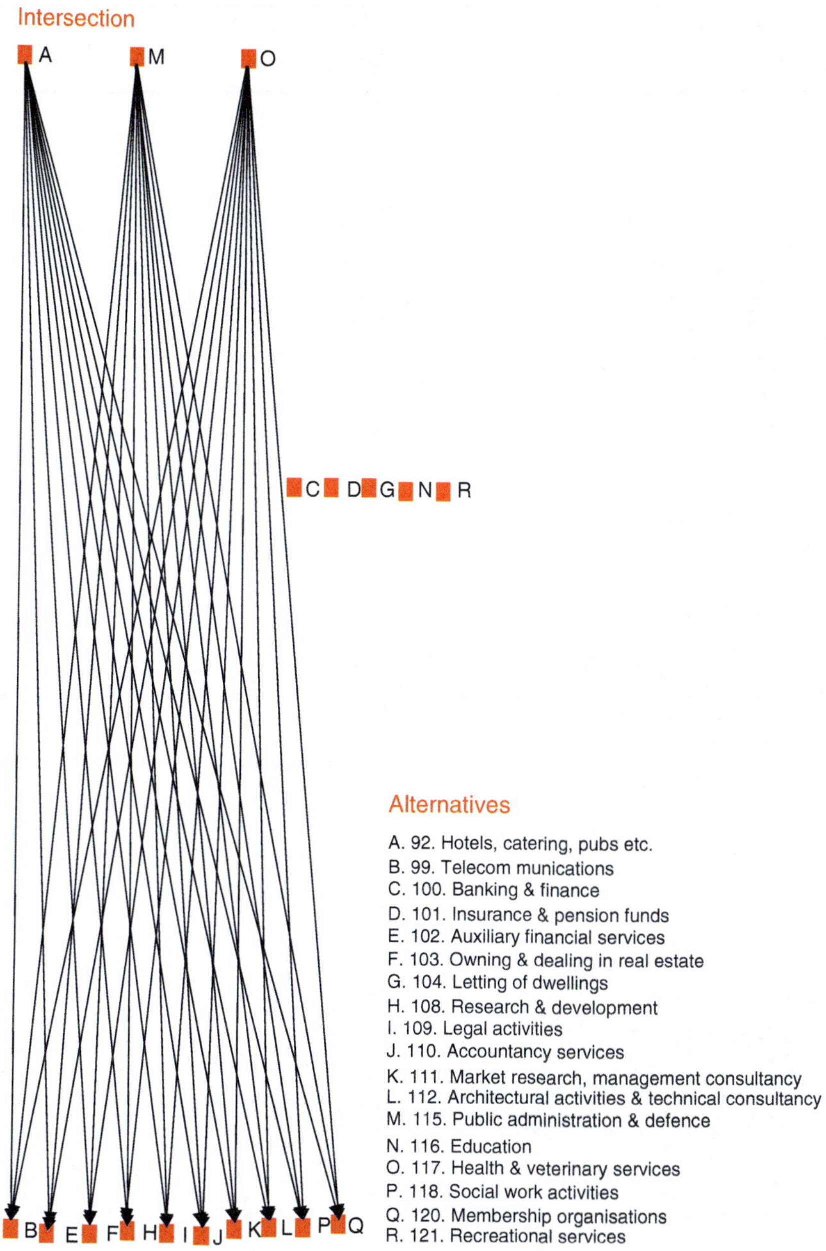

Fig. 6.12 The most sustainable sectors, UK, 2002, $\alpha=0.9$ – strong sustainability setting

Nomenclature of Economic Sectors, Input–Output Formulation, Office for National Statistics, UK, 2002

1 Agriculture
2 Forestry
3 Fishing
4 Coal extraction
5 Oil & gas extraction
6 Metal ores extraction
7 Other mining & quarrying
8 Meat processing
9 Fish & fruit processing
10 Oils & fats
11 Dairy products
12 Grain milling & starch
13 Animal feed
14 Bread, biscuits etc.
15 Sugar
16 Confectionery
17 Other food products
18 Alcoholic beverages
19 Soft drinks & mineral waters
20 Tobacco products
21 Textile fibres
22 Textile weaving
23 Textile finishing
24 Made-up textiles
25 Carpets & rugs
26 Other textiles
27 Knitted goods
28 Wearing apparel & fur products
29 Leather goods
30 Footwear
31 Wood & wood products
32 Pulp, paper & paperboard
33 Paper & paperboard products
34 Printing & publishing
35 Coke ovens, refined petroleum & nuclear fuel
36 Industrial gases & dyes
37 Inorganic chemicals
38 Organic chemicals
39 Fertilisers
40 Plastics & synthetic resins etc.
41 Pesticides

42 Paints, varnishes, printing ink etc.
43 Pharmaceuticals
44 Soap & toilet preparations
45 Other chemical products
46 Man-made fibres
47 Rubber products
48 Plastic products
49 Glass & glass products
50 Ceramic goods
51 Structural clay products
52 Cement, lime & plaster
53 Articles of concrete, stone etc.
54 Iron & steel
55 Non-ferrous metals
56 Metal castings
57 Structural metal products
58 Metal boilers & radiators
59 Metal forging, pressing etc.
60 Cutlery, tools etc.
61 Other metal products
62 Mechanical power equipment
63 General purpose machinery
64 Agricultural machinery
65 Machine tools
66 Special purpose machinery
67 Weapons & ammunition
68 Domestic appliances nec
69 Office machinery & computers
70 Electric motors & generators etc.
71 Insulated wire & cable
72 Electrical equipment nec
73 Electronic components
74 Transmitters for TV, radio & phone
75 Receivers for TV & radio
76 Medical & precision instruments
77 Motor vehicles
78 Shipbuilding & repair
79 Other transport equipment
80 Aircraft & spacecraft
81 Furniture
82 Jewellery & related products
83 Sports goods & toys
84 Miscellaneous manufacturing nec & recycling
85 Electricity production & distribution
86 Gas distribution

87 Water supply
88 Construction
89 Motor vehicle distribution & repair, automotive fuel retail
90 Wholesale distribution
91 Retail distribution
92 Hotels, catering, pubs etc.
93 Railway transport
94 Other land transport
95 Water transport
96 Air transport
97 Ancillary transport services
98 Postal & courier services
99 Telecommunications
100 Banking & finance
101 Insurance & pension funds
102 Auxiliary financial services
103 Owning & dealing in real estate
104 Letting of dwellings
105 Estate agent activities
106 Renting of machinery etc.
107 Computer services
108 Research & development
109 Legal activities
110 Accountancy services
111 Market research, management consultancy
112 Architectural activities & technical consultancy
113 Advertising
114 Other business services
115 Public administration & defence
116 Education
117 Health & veterinary services
118 Social work activities
119 Sewage & sanitary services
120 Membership organisations
121 Recreational services
122 Other service activities
123 Private households with employed persons

References

Anderson AW and Manning TW (1983) The Use of Input-Output Analysis in Evaluating Water Resource Development *Canadian Journal of Agricultural Economics* 31.1 15-26

Aroche-Reyes F (2003) A Qualitative Input-Output Method to Find Basic Economic Structures *Papers in Regional Science* 82.4 581-590

Ayres RU and Shapanka A (1976) Explicit Technological Substitution Forecasts in Long-Range Input-Output Models *Technological Forecasting and Social Change* 9.1–2 113-138

Barker T (1981) Projecting Economic Structure with a Large-Scale Econometric Model *Futures* 13.6 458-467

Barker T et al (1980) The Cambridge Multisectoral Dynamic Odel: An Instrument for National Economic Policy Analysis *Journal of Policy Modeling* 2.3 319-344

Barker T Ekins P and Foxon T (2007a) Macroeconomic Effects of Efficiency Policies for Energy-Intensive Industries: The Case of the UK Climate Change Agreements, 2000–2010 *Energy Economics* 29.4 760-778

Barker T, Junankar S et al (2007b) Carbon Leakage from Unilateral Environmental Tax Reforms in Europe, 1995–2005 *Energy Policy* 35.12 6281-6292

Cardenete MA and Sancho F (2006) Missing Links in Key Sector Analysis *Economic Systems Research* 18.3 319-325

Carter AP (1974) Energy, Environment, and Economic Growth *The Bell Journal of Economics and Management Science* 5.2 578-592

Carter AP (1976). *Energy and the Environment. A Structural Analysis* Waltham, Brandeis

Carter AP and Petri PA (1979) Aspects of a New World Development Strategy II: Factors Affecting the Long-Term Prospects of Developing Nations *Journal of Policy Modeling* 1.3 359-381

Common M and Stagl S (2005) *Ecological Economics: An Introduction*, Cambridge University Press

Daly H (2000) *Ecological Economics and the Ecology of Economics: Essays in Criticism* Cheltenham, Edward Elgar

Dietzenbacher E and Velázquez E (2007) Analysing Andalusian Virtual Water Trade in an Input–Output Framework *Regional Studies* 41.2 185-196

Duchin F (1986) Computers, Input-Output, and the Future *Journal of Economic Issues* 20.2 499-507

Duchin F (1990) The Conversion of Biological Materials and Wastes to Useful Products *Structural Change and Economic Dynamics* 1.2 243-261

Duchin F (1992) Industrial Input-Output Analysis: Implications for Industrial Ecology *Proceedings of the National Academy of Sciences of the United States of America* 89.3 851-855

Duchin F (1994) *Household Use and Disposal of Plastics an Input Output Case Study for New York City* New York, New York University

Duchin F (1998) *Structural Economics: Measuring Change in Technology, Lifestyles and the Environment* Washington DC, Island Press

Duchin F (2004) *Input-Output Economics and Material Flows* Troy, Rensselaer Polytechnic Institute

Duchin F and Szyld DB (1985) A Dynamic Input-Output Model with Assured Positive Output(*) *Metroeconomica* 37.3 269-282

Ferrer G and Ayres RU (2000) The Impact of Remanufacturing in the Economy *Ecological Economics* 32.3 413-429

Fontela E (1989) Industrial Structures and Economic Growth: An Input-Output Perspective *Economic Systems Research* 1.1 45-68

Forsund FR and Strom S (1976) The Generation of Residual Flows in Norway: An Input-Output Approach *Journal of Environmental Economics and Management* 3.2 129-141

Gay PW and Proops John LR (1993) Carbon Dioxide Production by the UK Economy: An Input-Output Assessment *Applied Energy* 44.2 113-130

Giljum S (2004) Trade, Materials Flows, and Economic Development in the South: The Example of Chile *Journal of Industrial Ecology* 8.1–2 241-261

Giljum S and Hubacek K (2004) Alternative Approaches of Physical Input–Output Analysis to Estimate Primary Material Inputs of Production and Consumption Activities *Economic Systems Research* 16.3 301-310

Graedel TE and Allenby BR (2002) *Industrial Ecology* Englewood Cliffs, Prentice Hall

Gutmanis I (1975) Input-Output Models in Economic and Environmental Policy Analyses *Proceedings of the IEEE* 63.3 431-437

Herendeen R and Tanaka J (1976) Energy Cost of Living *Energy* 1.2 165-178

Hewings GJD et al (1989) Key Sectors and Structural Change in the Brazilian Economy: A Comparison of Alternative Approaches and Their Policy Implications *Journal of Policy Modeling* 11.1 67-90

Hirschman AO (1958) *The Strategy of Economic Development* New Haven, Yale University Press

Hoekstra R (2005) *Economic Growth, Material Flows and the Environment. New Applications of Structural Decomposition Analysis and Physical Input-Output Tables* Cheltenham/ Northampton, Edward Elgar

Hoekstra R and van den Bergh JCJM (2002) Structural Decomposition Analysis of Physical Flows in the Economy *Environmental and Resource Economics* 23.3 357-378

Janssen R (1993) *Multiobjective Decision Support for Environmental Management* Dordrecht, Kluwer Academic Publishers

Jasny N (1962) The Russian Economic "Balance" and Input-Output Analysis: A Historical Comment *Soviet Studies* 14.1 75-80

Kananen I et al (1990) Multiple Objective Analysis of Input-Output Models for Emergency Management *Operations Research* 38.2 193-201

Kondo Y and Nakamura S (2005) Waste Input–Output Linear Programming Model with Its Application to Eco-Efficiency Analysis *Economic Systems Research* 17.4 393-408

Kossov VV (1964) Regional Input-Output Analysis in the U.S.S.R. *Papers in Regional Science* 14.1 175-181

Lantner R and Carluer F (2004) Spatial Dominance: A New Approach to the Estimation of Interconnectedness in Regional Input-Output Tables *The Annals of Regional Science* 38.3 451-467

Lenzen M (2003) Environmentally Important Paths, Linkages and Key Sectors in the Australian Economy *Structural Change and Economic Dynamics* 14.1 1-34

Lenzen M (2009) Understanding Virtual Water Flows: A Multiregion Input-Output Case Study of Victoria. *Water Resources Research* 45.9 W09416

Lenzen M and Foran B (2001) An Input-Output Analysis of Australian Water Usage *Water Policy* 3 321-340

Leontief W (1970) Environmental Repercussions and the Economic Structure: An Input-Output Approach *The Review of Economics and Statistics* 52.3 262-271

Leontief W (1974) Structure of the World Economy. Outline of a Simple Input-Output Formulation *The Swedish Journal of Economics* 76.4 387-401

Leontief W (1977a) Natural Resources, Environmental Disruption, and the Future World Economy. *Journal of International Affairs* 31.2 267

Leontief W (1977b) *The Future of the World Economy: A United Nations Study* New York, Oxford University Press

Leontief W and Duchin F (1986) *The Future Impact of Automation on Workers* New York, Oxford University Press

Leontief WW and Ford D (1972) Air Pollution and the Economic Structure: Empirical Results of Input-Output Computations *in* Brody A and Carter AP *Input-Output Techniques* Amsterdam, North-Holland

Luptáčik M and Böhm B (1994) An Environmental Input-Output Model with Multiple Criteria *Annals of Operations Research* 54.1 119-127

Madlener R and Stagl S (2005) Sustainability-Guided Promotion of Renewable Electricity Generation *Ecological Economics* 53.2 147-167

Moffatt I and Hanley N (2001) Modelling Sustainable Development: Systems Dynamic and Input-Ouput Approaches *Environmental Modelling and Software with Environment Data News* 16 545-557

Moran T, Angel M and del Rio Gonzalez P (2007) A Combined Input-Output and Sensitivity Analysis Approach to Analyse Sector Linkages and CO_2 Emissions *Energy Economics* 29.3 578-597

Munda G (1995) *Multicriteria Evaluation in a Fuzzy Environment* Heidelberg, Physica-Verlag

Munda G (2005) Multiple Criteria Decision Analysis and Sustainable Development *in* Figueira J, Salvatore G and Ehrgott M *Multiple-Criteria Decision Analysis. State of the Art Surveys* New York, Springer 953-986

Nakamura S (1999) An Interindustry Approach to Analyzing Economic and Environmental Effects of the Recycling of Waste *Ecological Economics* 28.1 133-145

Nakamura S and Kondo Y (2002) Recycling, Landfill Consumption, and CO_2 Emission: Analysis by Waste Input–Output Model *Journal of Material Cycles and Waste Management* 4.1 2-11

Nakamura S and Kondo Y (2006) A Waste Input-Output Life-Cycle Cost Analysis of the Recycling of End-of-Life Electrical Home Appliances *Ecological Economics* 57.3 494-506

Park S-H (1982) An Input-Output Framework for Analysing Energy Consumption *Energy Economics* 4.2 105-110

Peters GP and Hertwich EG (2006) Pollution Embodied in Trade: The Norwegian Case *Global Environmental Change* 16.4 379-387

Petri PA (1977) An Introduction to the Structure and Application of the United Nations World Model *Applied Mathematical Modelling* 1.5 261-267

Polenske KR and LinX (1993) Conserving Energy to Reduce Carbon Dioxide Emissions in China *Structural Change and Economic Dynamics* 4.2 249-265

Proops JLR (1977) Input-Output Analysis and Energy Intensities: A Comparison of Some Methodologies *Applied Mathematical Modelling* 1.4 181-186

Proops JLR (1984) Modelling the Energy-Output Ratio *Energy Economics* 6.1 47-51

Raa TT (1986) Applied Dynamic Input-Output with Distributed Activities *European Economic Review* 30.4 805-831

Rasmussen PN (1956) *Studies in Intersectoral Relations* Copenhagen, North-Holland

Rey G and Tilanus CB (1963) Input-Output Forecasts for the Netherlands, 1949–1958 *Econometrica* 31.3 454-463

Roy B (1985) *Methodologie multicritere d'aide a la decision* Paris, Economica

Schäfer D and Stahmer C (1989) Input–Output Model for the Analysis of Environmental Protection Activities *Economic Systems Research* 1.2 203-228

Shmelev SE and Giljum S (2004) Global Extraction of Renewable Resources: a Material Flows Analysis Perspective, Proceedings of the 8th Biennial Scientific Conference *"Challenging Boundaries: Economics, Ecology and Governance", International Society for Ecological Economics*, 11-14 July 2004, Montréal, Canada

Shmelev SE and Powell JR (2006) Ecological-Economic Modelling for Strategic Regional Waste Management Systems *Ecological Economics* 59.1 115-130

Shmelev SE and Rodríguez-Labajos B (2009) Dynamic Multidimensional Assessment of Sustainability at the Macro Level: The Case of Austria *Ecological Economics* 68.10 2560-2573

Simpson D and Tsukui J (1965) The Fundamental Structure of Input-Output Tables, An International Comparison *The Review of Economics and Statistics* 47.4 434-446

Söderbaum P (2000) *Ecological Economics: A Political Economics Approach to Environment and Development* London, Earthscan

Sonis M and Hewings GJD (1998) Economic Complexity as Network Complication: Multiregional Input-Output Structural Path Analysis *The Annals of Regional Science* 32.3 407-436

Sonis M and Hewings GJD (1999) Economic Landscapes: Multiplier Product Matrix Analysis Formultiregional IO Systems *Hitotsubashi Journal of Economics* 40 59-74

Sonis M et al (1995) Linkages, Key Sectors, and Structural Change: Some New Perspectives *The Developing Economies* 33.3 243-246

Stone R (1984) Model Design and Simulation *Economic Modelling* 1.1 3-23

Suh S (2005a) Developing a Sectoral Environmental Database for Input–Output Analysis: The Comprehensive Environmental Data Archive of the US *Economic Systems Research* 17.4 449-469

Suh S (2005b) Theory of Materials and Energy Flow Analysis in Ecology and Economics *Ecological Modelling* 189.3–4 251-269

Suh S (Ed) (2009) *Handbook of Input-Output Economics in Industrial Ecology* Dordrecht/London, Springer

Tukker A et al (2009) Towards a Global Multi-Regional Environmentally Extended Input-Output Database *Ecological Economics* 68.7 1928-1937

Vogt WG, Mickle MH and Aldermeshian H (1975) A Dynamic Leontief Model for a Productive System *Proceedings of the IEEE* 63.3 438-443

Wang H and Wang Y (2009) An Input–Output Analysis of Virtual Water Uses of the Three Economic Sectors in Beijing *Water International* 34.4 451-467

Wang L, MacLean HL and Adams BJ (2005) Water Resources Management in Beijing Using Economic Input-Output Modeling *Canadian Journal of Civil Engineering* 32 753-764

Chapter 7
Sustainable Development: Measuring Progress

Abstract This chapter focuses on the new *approach* to assessing progress towards sustainability based on the application of multicriteria decision-aid tools. Russia was chosen as a case study because of the 20 years of economic and social change, which have had a substantial impact on regional and sectoral patterns of the development of its economy, infrastructure, the quality of the environment, and the well-being of its people. The method is particularly suitable for countries in transition, where development has often been less than harmonious.

The chapter employs the UN Sustainable Development Framework of Indicators and assesses the sustainability of Russia using multi-criteria evaluation methods, namely the uncertainty randomization multi-criteria evaluation method "Analysis and Synthesis of Parameters under Information Deficiency" (ASPID). The analysis covers economic, environmental, and social trends in Russia's development between 1985 and 2007 and assesses the sustainability of this development from the point of view of multiple criteria.

The results show the potential of multi-criteria methods for sustainability assessment at the macro level and offer useful insights into the multidimensional nature of sustainability and the role of priority setting in the evaluation process. Such an analysis reveals the degree of harmony in sustainable development policy. It shows how different sets of priorities determine the outcome of multidimensional analysis of sustainability and might potentially help in assessing progress and designing new policy instruments. This chapter is one of the first to apply multi-criteria methods to macro sustainability analysis in a dynamic setting.

Keywords Sustainable development • Sustainability assessment • Macro-scale • Multi-criteria methods • Russia

S.E. Shmelev, *Ecological Economics: Sustainability in Practice*,
DOI 10.1007/978-94-007-1972-9_7,

Macro Sustainability Discussion

Sustainable development, understood here as harmonious development, considering environmental limitations, is essentially a multidimensional problem. It involves simultaneous analysis of environmental, economic, social and institutional aspects of the development of a state, a city or a region. The problem of sustainability on the macro scale has been addressed by many researchers: (Costanza and Patten 1995, Daly 1974, 1997, Daly and Cobb 1989, England 1998, Hanley et al. 1999, Lawn 2000, 2003, Max-Neef 1995, Neumayer 1999, 2000, Pearce and Atkinson 1993, Pearce et al. 1996). The alternative sustainability indicators, such as Human Development Index (HDI), Adjusted Net Savings (ANS) and Index of Sustainable Economic Welfare (ISEW) have been developed. Human Development Index (ul Haq 2003) is estimated for all countries of the world at the UN and is published in the Human Development Reports (UNDP 2009). The Adjusted Net Savings (Pearce and Atkinson 1993) is currently regularly calculated by the World Bank (Bolt et al. 2002). Adjusted net savings is the indicator of weak sustainability. The methodology of ISEW, developed by H. Daly and J. Cobb (Daly and Cobb 1989) has been applied to the UK (Jackson and Marks 1994), Sweden (Jackson and Stymne 1996), the Netherlands (Gerlagh et al. 2002), Austria (Stockhammer 1997) and other countries. The methodology of application of multi-criteria methods to the environmental problems has been developed by (Hovanov 1996, Janssen 1993, Larichev 1979, Munda 1995, Roy 1985) and other researchers. Recent applications of multicriteria methods to the analysis of sustainable development indicators can be found in (Munda 2005), etc.

Economic, social and environmental aspects of the development of Russia have been the focus of considerable research recently: (Glazyrina 2006, Granberg et al. 2002, Kalinichenko et al. 2007, Kuzyk and Yakovetz 2005, Lvov 2004, Reteyum 2004, Ryumina 2007). The history of sustainability analysis in Russia goes back to the works of Konstantin G. Gofman and his colleagues, who founded the Russian school of economics of nature management or "ecological economics" as it was sometimes called by Gofman (Gofman 1998, 2007). Sustainability analysis of specific sectors of the economy such as the energy sector, which is currently the key driving force of the Russian economy, has been undertaken in e.g. (Aslanyan et al. 2005), although social aspects of the development of the sector have been addressed only briefly. The current issues of sustainable development in Russia have also started to attract international attention (Oldfield 2001, Oldfield et al. 2003). However, there is still a gap in research on the comprehensive assessment of sustainability on the macro scale in Russia, interpretation of the links among the different social, economic and environmental processes and effects as well as strategic forward-looking analysis from the point of view of multiple criteria. A single priority of facilitating economic growth by doubling GDP is definitely limiting the sustainable development potential of the Russian economy.

It should be pointed out that, despite the value of single dimensional approaches to sustainability assessment – easy communication and use in policy making, there are fundamental problems, highlighted in (Martinez-Alier et al. 1998). Such problems include the issues of incomparability of values – can environmental and economic goods be substituted for one another in principle? What are the limits of such substitution? Have societies already reached these limits? Which production functions (Cobb-Douglas, CES type, etc.) should be used to describe most accurately the use of natural and economic factors of production? Which weighting should be used in such an assessment? There are also dynamic aspects to the problem. Discounting issues are a serious matter: can one discount environmental damages in principle? Is major damage done in the distant future considerably less important than damage done today? It seems we need to analyse the whole dynamic trajectory of development to be able to understand the dynamics of sustainable development. Some of these issues were addressed by (Shmelev and Rodríguez-Labajos 2009).

During recent decades, Russia has undergone dramatic structural, economic, social and institutional changes. These changes included the freeing of prices, the reviving of the entrepreneurship tradition, seizure of the previously substantial state support for science, attraction of direct foreign investment, development of the resource-extraction based economy, the relaxing of terms and conditions for international trade, first – dramatic deterioration and then a slow recovery in the level of consumption and quality of life, an introduction of a flat rate tax in 1997, which accelerated the growing differentiation between the rich and the poor. Joining the Kyoto Protocol in 2004, determining the emergence of government commitment on stabilising CO_2 emissions, record the high rates of economic growth in the past several years and declining life expectancy are additional brushstrokes in the complex picture of the development of the Russian economy.

In light of the above, it seems crucial to assess the progress of Russia towards sustainability by taking a "systems" or "holistic" perspective. This chapter will provide an overview of economic, environmental and social aspects of the development of Russia over the course of the past 20 years and will therefore analyse explicitly the sustainability of Russia's development. The multidimensional development path of Russia will be assessed with the help of multicriteria methods and an analysis of the complex trends and causes of unsustainability will follow. Application of multicriteria methods might help in the analysis of trade-offs among economic, environmental and social priorities.

The author will argue that a relative neglect of environmental and social aspects of the development of Russia has and continues to have long term sustainability consequences. The spatial aspect of the development of Russia presents another challenge, which has not been addressed adequately in the past.

Existing Approaches to Measuring Sustainability

First, the aggregate sustainability measures, such as HDI and ANS will be discussed; this will be followed by a detailed analysis of the economic, environmental and social aspects of Russia's development. It should be underlined that such aggregate methods assume that component indicators are perfect substitutes and large progress in one of them can compensate negative tendencies in many others. Such a peculiarity masks the existing multidimensional nature of the development process. It is for this reason that the author suggests new methods for the assessment of progress in the field of sustainable development, which are based on the application of multicriteria methods. The chapter will conclude with an application of multicriteria assessment tools and an analysis of multidimensional development trends.

Human Development Index

Human Development Index (HDI) is a composite measure, assessing achievements in the three main areas of human development: life expectancy, measured with the help of the life expectancy at birth index; education, measured with the help of the adult literacy index and good quality of life, measured with the help of real GPD per capita at PPP (ul Haq 2003). It should be noted that this paper uses statistical data from 1985 to 2007, supplied to the UNDP by the Russian Government; data on the component indices for 2006 was not available.

The tendencies of change in HDI in Russia were characterised by a substantial drop from 0,858 in 1991 to 0,804 in 1993 and a minimum of 0,747 in 1997. Starting from 1998, moderate growth in HDI in Russia is observed, and in 2007 its value reached 0,817 (this is the most recent data at the moment of publication of this chapter). It is instructive to see how changes in HDI are determined by changes in the indices of which it consists. The growth in HDI from 1998 was observed against a background of continuing decline in the life expectancy index; however, the dynamic growth in GDP and the moderate increase in the education index have led to the general change in trend and the positive dynamics of the Human Development Index. It should be noted that from the 28th place in the world in 1980 and 34th in 1990, Russia dropped to 52nd place as early as 1992 and reached an absolute minimum in 1995 (72nd place). In 1999, Russia occupied 55th place, in 2000 – 60th, in 2002 – 57th, but unfortunately in 2005 dropped to 67th and reached 71st in 2007. The following countries are slightly higher than Russia according to their level of development in 2007 – Albania, Belorussia, Rumania, Bulgaria, Malaysia; the following are slightly lower: Macedonia, Brazil, Columbia, Peru and Turkey. The position of Russia is considerably worse than that of Poland, Slovakia, Hungary, Lithuania, Argentina, Chile, Mexico, or Venezuela. The position of Russia is better than that of Ukraine, Georgia, Iran, Thailand, China, Jordan, Tunis, Gabon, Algeria, Indonesia, and Mongolia.

The Human Development Index of Russia for the period from 1985 to 2007 according to UNDP reports can be seen in Fig. 7.1.

It can be seen that full compensability between GDP, life expectancy and education determined the change in trend when growing GDP and education outweighed declining life expectancy. The complexity of the development pattern in HDI, therefore, was hidden in the linear aggregation procedure. If the incommensurability of values considered here (education, economic growth and life expectancy),

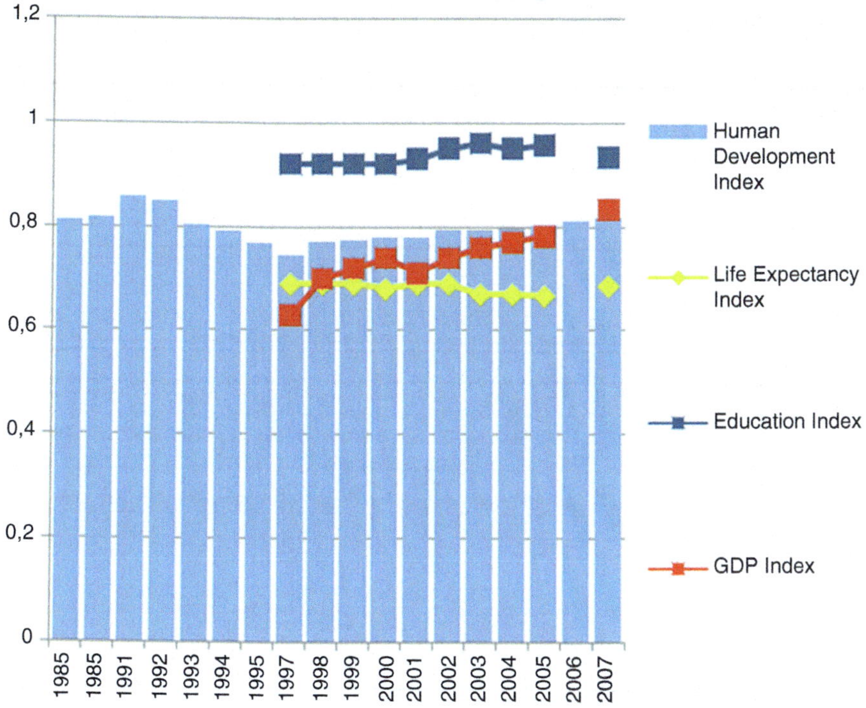

Fig. 7.1 Human development index in Russia, 1985–2007 and its constituent components (United Nations)

their different nature and different units of measurement are taken into account, the complexity of the development pattern is brought to light. The overall choice of the set of indicators becomes a crucial factor, since the number of criteria and the composition of the set will affect the assessment of the trend of development.

Adjusted Net Savings

Adjusted net savings, an indicator of "weak sustainability", denotes the level of capital which is accumulated within the economy less the depreciation of both produced and natural capital and environmental damage. "Weak" sustainability assumes that any type of capital is perfectly substitutable for natural capital as an input to production. From the adjusted net savings point of view, for example, a nation which reinvested all of its profits from the exploitation of non-renewable natural resources in the formation of human capital through its educational system would have imposed no net opportunity cost on future citizens of the country (Bolt et al. 2002).

As can be seen from Fig. 7.2, Adjusted Net Savings in Russia declined from 1995 to 1998; this was determined by the decline in gross national savings, an increase in the consumption of fixed capital, net forest-, energy- and mineral depletion and

Fig. 7.2 Adjusted net savings, Russia (Source: World Bank)

CO_2 damage. Since 2001, a tendency for the ANS to increase has been observed, but in 2006 its value has not risen above 0, which can be characterised as a struggle to minimise unsustainability levels. The most recent data, for 2008 characterise the situation as critical, but with some degree of hope. It is interesting to note that, as a result of recalculation the two issues of the World Bank ANS dataset are fundamentally different, which raises another concern regarding the reliability of such an estimate, especially in respect of to the fact that its constituent components are estimated in monetary terms.

If the issue of incommensurability of values is given equally serious consideration, it becomes apparent that increasing GDP and declining life expectancy cannot be considered equal substitutes. There is a clear need for development to be both beneficial for the economy and not destructive for the population and the environment. This requires the application of new methods of strategic policy analysis and decision making. The use of multi-criteria methods seems to be beneficial for such analysis for the following reasons: multi-criteria tools allow simultaneous consideration of several development objectives, interaction between decision makers and decision support systems allows one to gain deeper understanding of the links between different parts of the system and emergent properties of the system. Multi-criteria tools are capable of showing the trade-offs between often conflicting priorities and provide rankings of scenarios or alternatives based on multidimensional preference relationships.

The interested reader is invited to explore the paper (Shmelev and Rodríguez-Labajos 2009) published in the journal Ecological Economics, which explores these issues in greater depth, emphasising methods for the analysis of strong sustainability, where incommensurability of values is fully taken into account.

Spatial-Temporal Aspects of Development

In the following sections a spatial-temporal overview of the development of Russia over the past two decades will be given.

The trend in GDP growth has been seen by most observers as a positive tendency. The growth of the internal economy after the crisis of 1998 helped to overcome the consequences of the reforms and contributed to the alleviation of poverty. However, if one considers the structure of production for the range of years starting in 1990, it is possible to notice serious changes: a sharp increase in the share of wholesale and retail trade, and a decline in the share of agriculture and industry. At the same time, the growth of the informal economy has been observed, its share reaching 22–25% in 2000.

Let us consider the tendencies in the rate of domestic extraction of renewable and non-renewable resources in Russia (Shmelev and Giljum 2004). Domestic Extraction in Russia declined from 5.9 bln tones in 1992 to 4.3 bln tones in 1998, but has recovered since and reached 5.0 bln tones in 2002, which largely reflects the peculiarities of resource-extraction-led economic growth in Russia.

Atmospheric CO_2 emissions in Russia grew constantly between the end of the Second World War and 1980, with a mild decline in the rate of growth since 1980. From 1989 and, in particular, from 1990 to 1991 CO_2 emissions started to shrink, this was caused by the decline in production levels and structural changes in the economy. A historical minimum of the level of emissions in 1998 is comparable with the level of emissions between 1969 and 1970. Since 1999 emissions have started to grow again, but by the year 2002 they had not exceeded the values of emissions of 1996. As a whole, the existing tendency might be characterised as a positive one, however having declared the goals to double Russia's GDP without the proactive introduction of energy efficiency measures, and also gradual transition to renewable energy sources, Russia could face difficulties to meet its Kyoto protocol targets and subsequent commitments.

Social issues have been one of the most pressing problems for the Russian Federation over the past two decades. The dramatic fall in life expectancy (BMJ 1994) has been attributed to the increased incidence of heart disease, an increase in infant mortality and a rise in the numbers of deaths due to trauma. Of these three, the most severe rise has been in trauma, which includes industrial and road accidents, suicides, murders, military accidents and poisonings. Analysts link many of these to increased tension in society due to loss of jobs, restructuring of the economy and the difficult psychological climate in society.

The Gini Index of income inequality (measured for earnings) in Russia increased from 0.26 in 1991 to 0.409 in 1994 (larger values of the index correspond to the larger inequalities between rich and poor). After a brief decline to 0.375 in 1996, the Gini Index went up to 0.4 in 2003, reaching the value of 0.406 in 2004 and 0.423 in 2008. Therefore, in this dimension, Russia moved from the level of present day Austria, Luxembourg and Finland to the present day Moldova, Ukraine approaching the level of China, Turkey, the United States, and Uruguay.

The Unemployment rate in Russia climbed from 5.2% in 1992 to 13.3% in 1998 and then went down again to 7.8 in 2004. Inflation according to official data was always lower than that in Poland and approximately the same as in Ukraine.

The development of the Russian economy is characterised by extreme unevenness, if the spatial dimension is considered. The most prosperous regions are Moscow city, Moscow region, the oil and gas producing regions in the Urals and Siberia, and St Petersburg. The difference between the gross regional product in the most prosperous Moscow city and the less developed parts of Russia exceeds 100 times.

In the environmental dimension, spatial diversity is also considerable, with differences in total air emissions from stationary sources reaching 100 times, the level between certain regions.

Application of Multicriteria Methods

Taking the UN Sustainable Development Indicator Framework as a starting point, we have decided to apply a multicriteria assessment method to analyse the sustainability of the multidimensional development path of the Russian economy.

The Analysis and Synthesis of Parameters under Information Deficiency (ASPID) method, developed by Hovanov (Hovanov 1996) is based on the Bayesian model of uncertainty randomization. It is designed to compare complex objects, given a range of criteria describing their performance. To generate the set of weights used in the assessment, it takes into account non-numeric (ordinal) information on weight-coefficients values determined by a system $\mathbf{OI(w)=\{w_r=w_s;\ w_u>w_v;...\}}$ of equalities and inequalities for weight coefficients (indices **r, s, u, v** take values from set **{1,2,…,m}**); non-exact (interval) information on weight coefficient values determined by a system $\mathbf{II(w)=\{a_j<=w_j<=b_j;...\}}$ of inequalities and equalities (when $\mathbf{a_j=b_j}$) for weight-coefficients (index **j** takes values from set **{1,2,…,m}**) and non-complete expert knowledge. The final result of the assessment can be described as an ordering of analysed objects by estimated degrees of quality under evaluation (sustainability in our case). Therefore, within the framework of assessment, given the expressed priorities, the relationships of domination (in the sense of the chosen criteria set) emerge among the objects being assessed (the years of the country's performance in our case). The red and blue intercepts of a straight line, seen in the diagram, can be read in the following way: an abscissa of a midpoint of a red interval shows an average estimation of a correspondent object, while the interval's length is equal to the doubled standard deviation of the constructed aggregated preference index; an abscissa of a blue interval's right end shows the reliability for dominance relation between neighbouring aggregated estimations.

Table 7.1 Sustainable development criteria applied in the analysis of Russian economy (Based on the UN CSD Indicators of Sustainable Development 2007)

Theme	Sub-theme	Indicator
Poverty	Income inequality	Gini index of income inequality
Health	Mortality	Life expectancy at birth
Governance	Crime	Crimes per 100,000 inhabitants
Atmosphere	Climate change	Emissions of CO_2
Freshwater	Water quality	Water pollution, Nitrates
Economic development	Unemployment	Unemployment
Economic development	Macroeconomic performance	GDP per Capita
Economic development	Research and development	Expenditure on R&D as a GDP share
Consumption and production patterns	Energy use	Annual energy consumption per capita
Consumption and production patterns	Energy use	Share of consumption of renewable energy

The method was applied to two sets of 3 and 10 sustainability criteria over the same time period (1995–2003). Relational information on prioritisation of different criteria determined the weights, and as a result, randomised estimates of the domination of certain alternatives over others were obtained. The total list of criteria considered, based on the UN CSD Indicators of Sustainable Development (United Nations 2007) is presented in Table 7.1.

Dynamic Analysis

First, ASPID was applied in the case of three basic sustainability criteria: GDP per capita, CO_2 emissions and life expectancy, representing economic, environmental and social dimensions respectively (Figs. 7.3 and 7.4). The years from 1995 to 2003 were considered; this is represented on the vertical axis of the diagram. In each scenario assessed a set of assumptions was used to illustrate current policy priorities in the form of preference equalities and inequalities set.

First, the following priorities, reflecting the current policy trend, were set: GDP is more important than life expectancy, GDP is more important than reduction in CO_2 emissions, reduction of CO_2 emissions is more important than life expectancy. Such a set of priorities characterises the real development priorities obtaining in Russia.

It can be seen in Fig. 7.3 that the year 2006 dominates 2005, 2005 dominates 2004, 2004 dominates the year 2003 and so on, therefore an overall positive trend starting in 2000 can be seen. It should be underlined that this positive trend appears under specific conditions of relative importance of criteria, namely the priority of GDP over life expectancy and CO_2 emissions reductions and priority of CO_2 emissions reductions over life expectancy.

If, however, the different, more humanistic set of policy priorities is chosen as opposed to the more technocratic (Fig. 7.4), i.e. life expectancy is considered to be more important than GDP, and reduction in CO_2 emissions is seen as more important

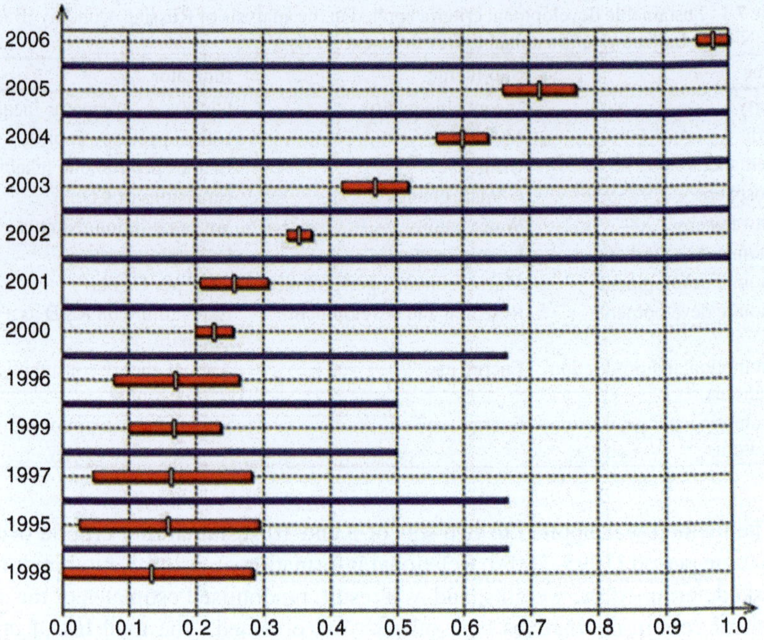

Fig. 7.3 Assessment results: 1995–2006, GDP per capita, CO_2 emissions, life expectancy; current policy priorities

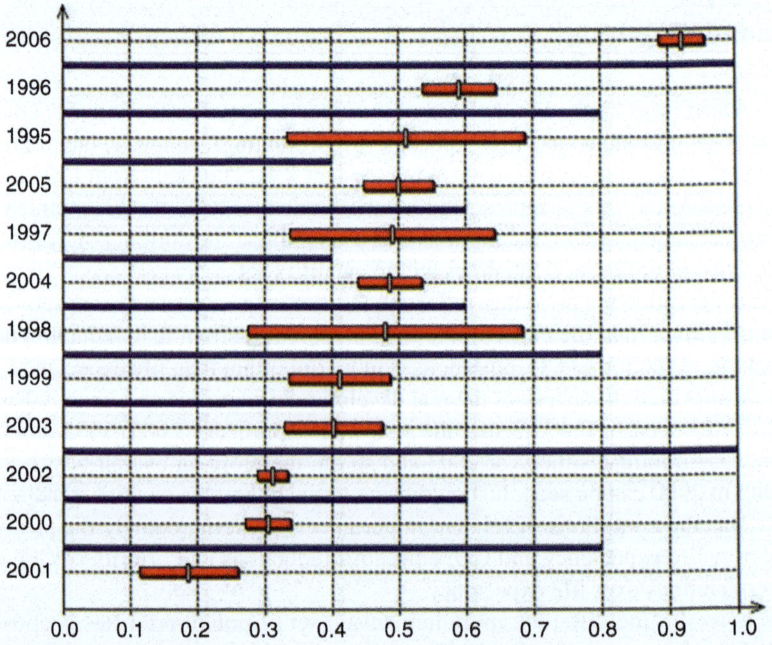

Fig. 7.4 Assessment results: 1995–2006, GDP per capita, CO_2 emissions, life expectancy; more humanistic policy priorities

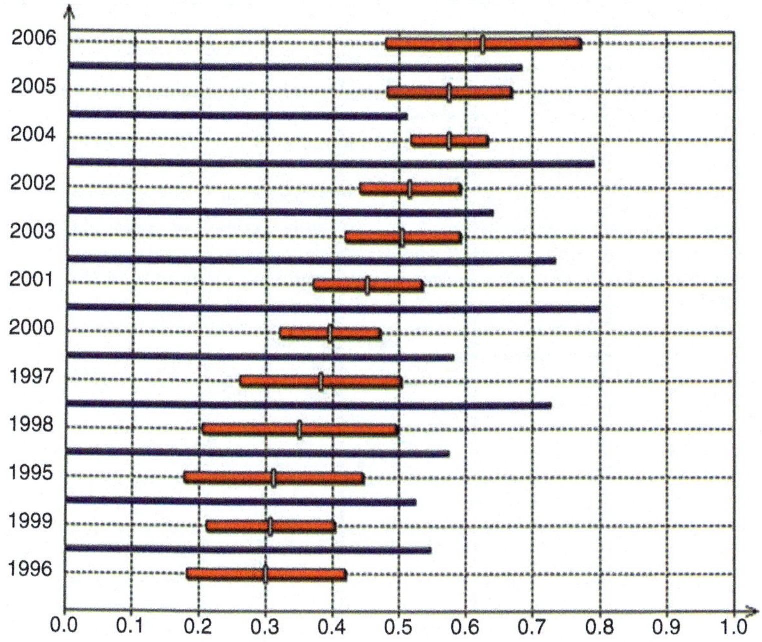

Fig. 7.5 Assessment results: 1995–2006, 10 criteria: current policy priorities

than GDP, then the trend is changing, and the most sustainable years in this setting are 2006, followed by 1996 and 1995, then 2005, then 1997, then 2004, then 1998 and so on. The least sustainable years in this setting were 2001, 2000, 2002, 2003 and 1999.

In the more detailed analysis the following ten criteria were taken into account: economic – GDP per capita, Total Primary Energy Supply (TPES) per capita, share of renewables; environmental: CO_2 emissions, water pollution; social: life expectancy, GINI index of income inequality, unemployment rate, crimes per 100,000 inhabitants; and institutional: investment in R&D.

The first case (Fig. 7.5), illustrates a current policy priority scenario: GDP growth is more important than life expectancy and CO_2 emissions. As can be seen from Fig. 7.5, given the assumptions above, the "sustainability trend" appears to be positive up until 2006 (with minor exceptions), with more recent years dominating the previous years.

If, however, a different pro-environmental and more humanistic set of policy priorities is assumed – an increase in life expectancy and reduction in CO_2 emissions to combat climate change are more important than GDP growth, etc. the picture becomes quite different (Fig. 7.6). And now the years 1997 and 1998 dominate the other years and since 1998 a decline in sustainable well-being is observed. The years 2005, 2006 and 1995 appear to be the least sustainable in this setting. It should be noted that due to the larger number of criteria in the extended set, the uncertainties in domination, represented by the length of the red lines around the probabilities, are considerably higher.

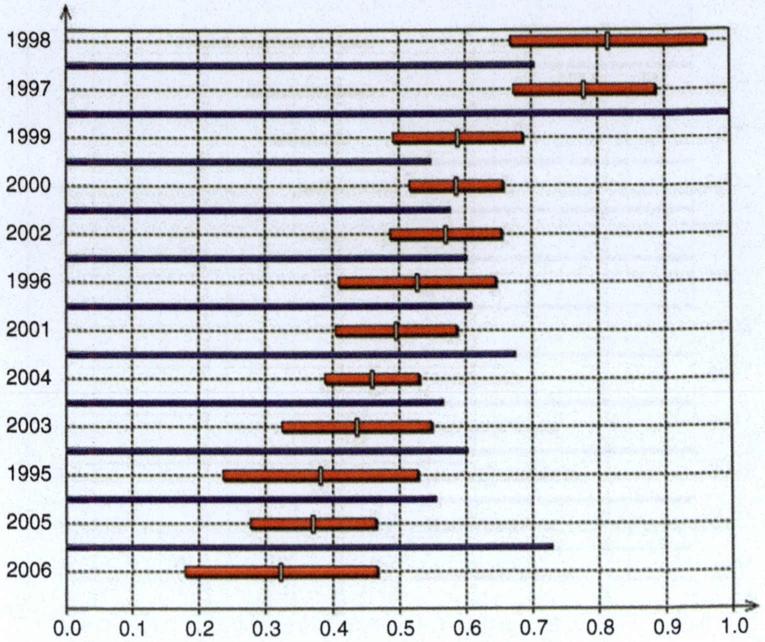

Fig. 7.6 1995–2006, 10 criteria: more humanistic policy priorities

Spatial Setting

Spatial data presents another important aspect of sustainability analysis in the Russian context. Large disparities between Russian regions in the value of gross regional product, life expectancy and CO_2 emissions make spatial sustainability assessment an interesting and worthwhile exercise. We will show here two major results which correspond to the priorities set in the dynamic assessment: emphasis on life expectancy and emphasis on economic output. It is interesting to note that if life expectancy is taken as a primary sustainability criterion, the first ten most sustainable regions become: Ingushetia, Dagestan, Moscow, Kabardino-Balkaria, Karachaevo-Cherkessia, Severnaya Ossetia, Belgorod Oblast, Adygeya, Stavropol Oblast and Krasnodarsky Kraj. If economic output is taken as the most important criteria, then Tyumen region (where a large proportion of the natural resources is mined), Moscow (where the taxes are collected), followed by Tatarstan, Lipetskaya Oblast, Ingushetia, Tomskaya Oblast, St Petersburg, Omskaya Oblast, Belgorod Oblast and Dagestan become the leading regions in terms of sustainability. The results of the analysis allow us to conclude that the development of Russian regions is characterised by extreme unevenness. Depending on the chosen set of priorities, completely different regions appear as more sustainable in the rating.

In light of these facts, it seems extremely important and desirable to undertake similar regional assessments using larger sets of criteria including crime rates, income differentiation, unemployment, emissions of substances other than CO_2, resource use, generation of waste, consumption of energy, the share of renewables in the energy mix and other sustainability indicators.

The methodology presented here could help to highlight regional problems which need to be addressed and to find the regional sustainability leaders which need to be supported. It should be underlined that such application of multicriteria methods for the analysis of dynamic tendencies of sustainable development in the Russian context is undertaken for the first time.

Discussion

Treatment of many conflicting priorities simultaneously is a challenge which many national governments and international organisations face today.

We have seen that the positive trend in the Human Development Index, coupled with the increasing, and then decreasing values of Adjusted Net Savings, as well as trends in the different additional sustainability criteria based on the United Nations sustainable development indicators framework create a multifaceted picture of the development of Russian society, further complicated by enormous regional disparities.

Specific policy priorities, as was shown in this chapter, can determine the result of the evaluation of "progress", the interpretation of which rests heavily on social consensus and shared values. We have seen that placing more emphasis on social aspects of development, such as longer and healthier life and reduction of income inequalities, as well as environmental aspects, such as cleaner air, climate change mitigation, increased deployment of renewable energy technologies, and contribution towards global sustainability as opposed to increase in the GDP, change the interpretation of the progress that society experienced in a particular time frame. Therefore, the hierarchy of policy priorities which are supported by the given society or international community can stimulate a pattern of more or less sustainable development.

The solution for the current critical situation seems to be the following – growth in education expenditure, increase in government and stimulation of private investment in the national economy; the use of cleaner technologies (minimization of CO_2 emissions), a transition to more extensive use of renewable energy (minimisation of natural capital depletion in the long run), and more efficient use of energy in different sectors, development of sustainable waste management systems, capable of returning valuable resources to economic circulation and thereby reducing environmental impacts.

It can also be seen in the assessment that increasing number of criteria bringing relevant dimensions into the evaluation framework further increases the degree of uncertainty of domination of particular periods of assessment over others, which is depicted in the length of the bars around the probabilities of domination in respective charts. The application of multicriteria assessment methods, therefore, can be a valuable tool for policy analysis and can help deal with high levels of complexity in a sustainability assessment problem. Such assessments can stimulate debate on the nature of sustainability and the vector of development of particular countries or regions and can improve understanding of the links between the constituent parts of the multidimensional evolving economy-society-environmental system.

Thus, the proposed approach offers a comprehensive framework for the assessment of sustainability at the macro level and could provide necessary support for policy makers in establishing priorities for development as well as evaluation of progress in a multi-dimensional setting. In the context of the evolving economy of Russia, it seems that more emphasis is needed on the elicitating of social preferences and democratic articulation of different interests within a society, so that social and environmental issues can become as important as the speed of economic development; and the true sustainability of development could be secured. The proposed model also illustrates the need to conduct active policy in these fields, which are areas of relative "unsustainability" in Russia. Additional measures to reduce the gap between rich and poor should be undertaken, for example with the help of a progressive taxation system; active government investment in the sciences should support and develop research potential, additional investment should be directed towards the development of the health care system, the development of environmental management systems, including the preservation of forests, development of the waste management systems, development of renewable energy systems, as well as creation of an environment, capable of securing an increase in life expectancy. We would like to hope that Russia could achieve more progress in the field of sustainable development.

References

Aslanyan G, Molodsov S and Iacobtchouk V (2005) Monitoring the Sustainability of Russia's Energy Development *Natural Resources Forum* 29 334-342

BMJ (1994) Life expectancy in Russia falls. Available at: http://www.bmj.com/cgi/content/full/308/6928/553

Bolt K, Clemens M and Matete M (2002) *Manual for Calculating Adjusted Net Savings* Washington DC, Environment Department, World Bank/Washinton DC

Costanza R and Patten BC (1995) Defining and Predicting Sustainability *Ecological Economics* 15.3 193-196

Daly HE (1974) Steady-State Economics Versus Growthmania: A Critique of the Orthodox Conceptions of Growth, Wants, Scarcity, and Efficiency *Policy Sciences* 5.2 149-167

Daly H (1997) *Beyond Growth: The Economics of Sustainable Development* Boston, Beacon Press

Daly HE and Cobb J (1989) *For the Common Good: Redirecting the Economy Toward Community, the Environment, and a Sustainable Future* Boston, Beacon Press

England RW (1998) Measurement of Social Well-Being: Alternatives to Gross Domestic Product *Ecological Economics* 25.1 89-103

Gerlagh R et al (2002) A Measure of Sustainable National Income for the Netherlands *Ecological Economics* 41.1 157-174

Glazyrina IP (2006) Issledovanie kachestva rosta regionalnoj ekonomiki v kontekste konzepzii ustojchivogo razvitija *Ekonomika prirodopolzovanija* 4 21-31

Gofman KG (1998) *Ekonomika prirodopolzovanija (iz nauchnogo nasledija)* Moscow, URSS
Gofman KG (2007) *The Work of Konstantin G. Gofman and Colleagues: An Early Example of Material Flow Analysis from the Soviet Union* Vienna, IFF
Granberg AG et al (2002) *Strategija i problemy ustojchivogo razvitija Rossii v XXI veke* Moscow, Economica
Hanley N et al (1999) Measuring Sustainability: A Time Series of Alternative Indicators for Scotland *Ecological Economics* 28.1 55-73
Hovanov NV (1996) *Analiz i sintez pokazatelej pri informazionnom defizite* St Petersburg, St Petersburg University Press
Jackson T and Marks N (1994) *Measuring Sustainable Economic Welfare: A Pilot Index UK 1950–1990* London, Stockholm Environment Institute, Stockholm
Jackson T and Stymne S (1996) *Sustainable Economic Welfare in Sweden: A Pilot Index 1950–1992* Stockholm Environment Institute, Stockholm
Janssen R (1993) *Multiobjective Decision Support for Environmental Management* Dordrecht, Kluwer Academic Publishers
Kalinichenko VV, Surovtsov LK and Shalabin GV (2007) Analiz problem ekonomicheskogo rosta Rossii s ispolzovanijem dinamicheskoj mnogootraslevoj modeli Nejmana *Vestnik SPbGU, Seria* 5.3 13-25
Kuzyk BN and Yakovetz YV (2005) *Rossiya 2050 in Strategija Innovazionnogo Proryva* (2nd edition) Moscow, Economica
Larichev OI (1979) *Nauka i isskusstvo prinyatija reshenij* Moscow, Nauka
Lawn P (2000) *Toward Sustainable Development: An Ecological Economics Approach* Boca Raton, Lewis Publishers/International Society for Ecological Economics
Lawn PA (2003) A Theoretical Foundation to Support the Index of Sustainable Economic Welfare (ISEW), Genuine Progress Indicator (GPI), and Other Related Indexes *Ecological Economics* 44.1 105-118
Lvov DS (2004) *Rossija v globalizirujuschemsya mire: Politico-economicheskije ocherki* Moscow, Nauka
Martinez-Alier J, Munda G. and O'Neil J (1998) Weak Comparability of Values as a Foundation for Ecological Economics *Ecological Economics* 26.3 277-286
Max-Neef M (1995) Economic Growth and Quality of Life: A Threshold Hypothesis *Ecological Economics* 15.2 115-118
Munda G (1995) *Multicriteria Evaluation in a Fuzzy Environment* Heidelberg, Springer
Munda G (2005) "Measuring Sustainability": A Multi-Criterion Framework *Environment, Development and Sustainability* 7.1 117-134
Neumayer E (1999) *Weak Versus Strong Sustainability: Exploring the Limits of Two Opposing Paradigms* Cheltenham/Northampton, E. Elgar
Neumayer E (2000) On the Methodology of ISEW, GPI and Related Measures: Some Constructive Suggestions and Some Doubt on the "Threshold" Hypothesis *Ecological Economics* 34.3 347-361
Oldfield J (2001) Russia, Systemic Transformation and the Concept of Sustainable Development *Environmental Politics* 10.3 94-110
Oldfield J, Kouzmina A and Shaw D (2003) Russia's Involvement in the International Environmental Process: A Research Report *Eurasian Geography and Economics* 44.2 157-168
Pearce DW and Atkinson GD (1993) Capital Theory and the Measurement of Sustainable Development: An indicator of "weak" Sustainability *Ecological Economics* 8.2103-108
Pearce D, Hamilton K and Atkinson G (1996) Measuring Sustainable Development: Progress on Indicators *Environment and Development Economics* 1.01 85-101
Reteyum AY (2004) *Dvenadzat' let iz zhizni strany* Moscow, Chorion
Roy B (1985) *Méthodologie Multicritère d'Aide à la Décision* Economica, Paris
Ryumina EV (2007) Kriterii syrjevoj zavisimosti ekonomiki regionov Rossii. Est' li u Rossiinesyrjevoje buduschee? In *Tezysy vystuplenij uchastnikov VI Mezhdunarodnoj Kondratjevskoj konferenzii. Mezhdunarodnyj Fond N.D. Kondratjeva* Moscow 61-63
Shmelev SE and Giljum S (2004) Global Extraction of Renewable Resources: A Material Flows Analysis Perspective. *Proceedings of the 8th Biennial Scientific Conference "Challenging*

Boundaries: Economics, Ecology and Governance", International Society for Ecological Economics, July 11–14, 2004, Montréal, Canada

Shmelev SE and Rodríguez-Labajos B (2009) Dynamic Multidimensional Assessment of Sustainability at the Macro Level: The Case of Austria *Ecological Economics* 68.10 2560-2573

Stockhammer E (1997) The Index of Sustainable Economic Welfare (ISEW) as an Alternative to GDP in Measuring Economic Welfare. The Results of the Austrian (revised) ISEW calculation 1955–1992 *Ecological Economics* 21.1 19-34

ul Haq M (2003) The Birth of the Human Development Index *in Readings in Human Development* Oxford, Oxford University Press 127-137

UNDP (2009) Human development report, overcoming barriers *Human Mobility and Development* Available at: http://hdr.undp.org/en/reports/

United Nations (2007) *Indicators of Sustainable Development, Guidelines and Methodologies* (3rd edition) New York, United Nations

Part II
Ecological-Economic Applications

Chapter 8
Climate Change and Renewable Energy: How to Choose the Optimal Pool of Technologies

Abstract Renewable energy is seen as an alternative to conventional energy sources, which are predominantly fossil-fuel based and cause dangerous climate change. Various forms of renewable energy also require some resource inputs and installations need to be replaced every 30 years or so. This leads an analyst to the problem of careful selection of a pool of energy sources. Although the conventional form of analysis has largely been centred on cost-minimization, more and more additional criteria, belonging to the economic, social, technical, risk, resource and emissions realms are introduced to make justified decisions on the selection of one technology or of one combination over the other. This chapter makes the first step towards this goal by providing taxonomy of criteria used in decision making regarding renewable energy; it also reviews existing modelling approaches which are currently used in the field. Methodologically, the emphasis in this chapter is on multicriteria decision aid tools.

Keywords Renewable energy • Decision making • MCDA • Sustainability • Ecological-economic modelling

The Energy System

The energy system has been the subject of substantial discussion over the course of the last 40 years or so, but each time the discussion intensified, it was for a different reason. First, there was an oil crisis and everyone was concerned about energy security, "peak oil" and high oil prices, later on the focus shifted toward climate change, with the energy system being the largest contributor of anthropogenic GHG emissions. Various alternatives have been considered: renewable energy, so called carbon capture and storage, energy demand reduction. Experiments with reality are usually very costly and could lead to undesired consequences. It is for this reason that mathematical models have been employed to help understand the functioning of the energy system, make justified decisions about its development, choose the best

S.E. Shmelev, *Ecological Economics: Sustainability in Practice*,
DOI 10.1007/978-94-007-1972-9_8, © Springer Science+Business Media B.V. 2012

technologies for combination in a particular region, or design a strategy for CO_2 emissions reduction of 80% over the course of the following 40 years. The energy system is complex, with the whole being more than the sum of its parts.

Because various elements of the system and different decisions need to be modelled, several approaches have been developed which, taken separately or in combination, help to improve our understanding of the energy system. It should be noted that modelling should not be aimed at providing a final answer, but its main purpose is to explore possibilities, learning about the complexities of the system, which often come from the interaction between the analyst or decision maker and the model.

Methods

The most frequently used approaches to the modelling of the energy system have been: optimization; input–output analysis (IO), life cycle assessment (LCA) and multicriteria decision aid (MCDA).

Optimization is a tool invented during the Second World War to plan the effectiveness of military and supply operations. Its main aim is to find the solution (a vector in a multidimensional space) which maximises or minimises a certain function, subject to certain constraints. Various structures of problems (among the standard formulations are such exotic types as "the travelling salesman problem", "the knapsack problem", or "the diet problem") determine the most adequate mathematical formulation. The options available to the analyst are: continuous or integer variables denoting decisions; the shape of the goal function – linear, concave, quadratic; the use of time in the model and dependence of future states of the system on past decisions (dynamic programming); the degree of uncertainty associated with problem parameters (interval programming, stochastic programming) or the number of explicit goal functions (single criteria or multiple criteria programming).

Historically, Stefan Rath-Nagel and Kenneth Stocks in 1981 developed a multicriteria optimization problem that described the functioning of the energy management system, had a detailed technological choice component and was designed to help in planning over the relatively long periods of time. The most prominent feature of the article was a "Price/Security" trade-off curve. Later on this model was developed further at IEA and became a well known MARKAL model. It is now used by all OECD governments to plan their energy systems and draw plans for CO_2 emissions reductions. Unfortunately for the energy system and the development of modelling techniques, the idea of a multiple criteria optimization problem was abandoned, and the MARKAL model was reduced to the cost-minimization tool. It should be said that usually a wealth of aspects of the energy system need to be taken into account: economic, social, technical, environmental; and to reduce such complexity to a single-criterion is gravely to oversimplify the real process.

Another major method, input–output analysis is used to analyse what will happen to the whole economy if certain changes are made in one part of it (i.e. energy system). The method was designed by the Nobel Prize winning economist W. Leontief and is based on a system of linear equations which link intermediate inputs to production processes with final demand. The national economy in this case is considered as a network of sectors, homogeneous types of economic activity such as car manufacturing, aviation or education and health care. The beauty of this method comes from the fact that not only direct but also indirect economic impacts can be traced. Environmental extensions which exist allow simultaneous consideration of economic and environmental impacts, which could take resource use, emissions of various substances as well as the use of water into account.

Life cycle assessment is a method for the detailed description of various processes from the point of view of inputs and outputs at each stage of the process. A system or a process is usually broken down into elementary stages, and the use of energy and materials at each stage is meticulously reordered. When all the data is collected, aggregate indices of impact are computed, representing global warming potential, acidification, toxicity and so on. Such representation allows a comparison between certain stages of the process, and is usually aimed at testing alternative technological solutions or materials. In the energy sector, many applications are focused on individual technologies, e.g. life cycle analysis of nuclear energy, wind turbines or solar PV installations.

Multicriteria decision aid is a group of methods rooted in the works of the French mathematician Prof. Bernard Roy. They are often used when it is necessary to compare several technologies or scenarios from the point of view of a whole spectrum of impacts, e.g. economic (installation and management costs), social (employment, effects on community), environmental (CO_2 and SO_2 emissions), resource use, etc. Several MCDA tools became more prominent: ELECTRE, PROMETHEE, NAIADE. Discrete methods are usually designed to compare several alternatives, whereas multicriteria optimization may work with very large spaces of alternatives.

There is a substantial literature dealing with the application of multicriteria decision aid (MCDA) tools to planning and investment in energy alternatives. Most studies were published after 1990. They use a diverse set of MCDA methods, including AHP (Chatzimouratidis and Pilavachi 2008, Noble 2004); ASPID (Afgan and Carvalho 2002); MACBETH (Burton and Hubacek 2007); ELECTRE (Georgopoulou et al. 1997, Mavrotas et al. 2003, Siskos and Hubert 1983); PROMETHEE (Diakoulaki and Karangelis 2007, Haralambopoulos and Polatidis 2003, Madlener and Stagl 2005) and NAIADE (Cavallaro and Ciraolo 2005, Gamboa and Munda 2007).

The first important study is Siskos and Hubert (1983), who dealt with the comparison of energy alternatives in the context of France from a social and public health point of view. Six major energy systems were compared: oil, coal, nuclear, two types of solar thermal and solar photovoltaic. The ELECTRE III MCDA method was used to compare these alternative options where the following set of criteria was employed: accidents, public risk, individual risk, collective risk, cost of kWh, work content, balance of payments, creation of jobs, available resources, securing supplies, and technical feasibility.

Georgopoulou et al. (1997) employed ELECTRE III to study the choice among alternative energy policies for the Greek island of Crete. The researchers emphasise the multicriteria nature of the strategic problem at hand and criticise dominant cost-benefit approaches. The criteria identified include: investment costs, operation and maintenance cost, reliability in covering peak demand, operationality, stability of the network, cohesion to local activities, regional employment, air quality, noise, visual disamenity, depletion of finite energy resources, risk of climate change, ecosystems protection, land use, implementation of EU environmental policy.

Afgan and Carvalho (2002) use the ASPID (Analysis and Synthesis of Parameters under the Information Deficiency) MCDA method to compare the following technologies: coal, solar thermal, geothermal, biomass, nuclear, PV solar, wind, ocean, hydro, and gas using a set of five sustainability criteria: efficiency, installation cost, electricity cost, CO_2 emissions and area required.

Haralambopoulos and Polatidis (2003) employ the PROMETHEE II MCDA tool to justify group decision making on the development of geothermal technology in the Greek island of Chios. The following five criteria were taken into account: conventional energy saved (toe/year), return of investment (yearly earnings per initial investment) and number of jobs created, environmental pressures and entrepreneurial risk of investment.

Mavrotas et al. (2003) apply a combination of the ELECTRE TRI approach with integer linear programming to select the best applications for wind energy development in Greece. Such an interaction of methods allows us to generate different combinations of structural parameters of the problem as well as to carry out a grouping of alternatives when no strict differentiation among alternatives is required (ELECTRE TRI is capable of assigning a group of objects to one of the predefined classes).

Noble (2004) assesses five development scenarios for the Canadian energy system given the following criteria: atmospheric emissions, resource efficiency, energy security, economic factors, public health and safety, etc. Following the Delphi method to extract expert opinions, an Analytical Hierarchy Process (AHP) method is applied to perform multicriteria evaluations. At the national level, the assessment panel identified alternative A3, which emphasises an increase in renewable energies, electricity diversification and improvements in fossil-fuel technologies as the preferred option for Canada's electricity future. Stakeholder and group preference analysis is carried out as well.

Cavallaro and Ciraolo (2005) employ a multicriteria assessment using the NAIADE method to evaluate the feasibility of installing wind turbines on the Italian island of Salina. Four different scenarios are considered, varying in term of capacity and number of installations, using the following criteria: investment cost, operating and maintenance costs, energy production capacity, fuel savings, technological maturity, realisation times, CO_2 emissions avoided, visual impact, acoustic impact, impact on ecosystems, social acceptability.

Madlener and Stagl (2005) propose a comprehensive methodology for the assessment of renewable energy technologies using a structured set of criteria. The set is composed of a range of indicators, representing a biophysical dimension: Resource inputs needed for production (land resources, water, material requirements, indirect energy requirements), potential environmental consequences (impacts on natural biota, habitats and wildlife, environmental risks, visual intrusion, impact on microclimate, impact on soil productivity, impact of resettlements), potential consequences of energy conversion and use (air pollution, organic emissions, generation of solid wastes, water pollution, pressure on land and water resources and other hazards), and socio-economic impacts (employment, occupational hazards, noise, impact on local poverty, household income disparity, democratic control over markets, safety of power supply, impact on balance of trade, long-term economic viability, local net value added, economic risk to ratepayers, impact on flexibility of supply). The authors suggest that Promethee II be used as a MCDA tool for this type of problem.

Gamboa and Munda (2007) explore the problem of wind farm location in Catalonia, Spain using the NAIADA MCDA approach. The following criteria are taken into account: land owner's income, distribution of income, income of municipalities, number of jobs, visual impact, forest loss, noise annoyance, avoided CO_2 emissions, and installed capacity. Stakeholder analysis is performed to understand how stakeholder coalitions might be formed.

Burton and Hubacek (2007) study the implementation of renewable energy schemes in a local borough of the County of Yorkshire with the help of the MACBETH method. The following technologies are compared: solar PV, micro-hydro, micro-wind, biomass, large scale wind, landfill gas, large scale hydro, energy from waste. The criteria taken into account are: capital cost, operation and maintenance, generation capacity, lifespan, carbon emissions, noise, natural environment and social consequences.

Diakoulaki and Karangelis (2007) apply the PROMETHEE method in order to compare several energy strategies for Greece using the following criteria: investment cost, production cost, guaranteed energy, available power during peak load, security of supply, CO_2 increase, SO_2 emissions, and NOx emissions.

Chatzimouratidis and Pilavachi (2008) evaluate ten energy generation technologies: coal, oil, natural gas turbine, natural gas combined cycle, nuclear, hydro, wind, photovoltaic, biomass, and geothermal using the Analytical Hierarchy Process. The following criteria are taken into account: quality of life (accident fatalities, NMVOCs, CO_2-eq, NOx, SO_2, PM and land required) and socio-economic aspects (job creation, compensation rates, social acceptance).

All the approaches outlined above are summarised in Table 8.1, which is an important structural element for extracting the taxonomy of criteria for sustainable energy systems decision making. One can see that a wide spectrum of methods have been used to support energy decisions, and although some papers have been focused on the comparison of individual technologies, others looked at the comparison of strategies, including deployment of multiple types of energy technologies.

Taxonomy of Criteria

Figure 8.1 presents a the taxonomy of sustainable energy systems decision criteria. They are broadly differentiated between six major groups: economic (C), social (S), resource inputs (R), emissions (E), risks (K) and technical feasibility (T). Each group is represented by a range of criteria found in the literature.

Table 8.1 MCDA methods in sustainable energy research

Authors and year	Application	MCDA method	Criteria	Alternatives
Siskos and Hubert (1983)	France	ELECTRE III	Accidents, public risk, individual risk, collective risk, cost of kWh, work content, balance of payments, creation of jobs, available resources, securing supplies, and technical feasibility	Oil, coal, nuclear, two types of solar thermal and solar photovoltaic
Georgo-poulou et al. (1997)	Crete, Greece	ELECTRE III	Investment costs, operation and maintenance cost, safety in covering peak demand, operationality stability of the network, cohesion to local activities, regional employment, air quality, noise, visual disamenity, depletion of finite energy resources, risk of climate change, ecosystems protection, land use, implementation of EU environmental policy	Energy policies
Afgan and Carvalho (2002)		ASPID	Efficiency, installation cost electricity cost, CO_2 emissions and area required	Coal, solar thermal, geothermal, biomass, nuclear, PV solar, wind, ocean, hydro, and gas
Haralambopoulos and Polatidis (2003)	Chios, Greece	PROMETHEE II	Conventional energy saved (toe/year), return of investment (yearly earnings per initial investment) and number of jobs created, environmental pressures and risk of entrepreneurial investment	Geothermal energy
Mavrotas et al. (2003)	Greece	ELECTRE TRI + MILP	Internal Rate of Return (IRR) of the investment (30), maturity of the certification procedure (MCP) (31), and quality of application (AQ) (32)	Wind energy development applications
Noble (2004)	Canada	AHP	Atmospheric emissions, resource efficiency, energy security, economic factors, public health and safety etc.	Energy strategies

Cavallaro and Ciraolo (2005)	Salina, Italy	NAIADE	Investment cost, operating and maintenance costs, energy production capacity, fuel savings, technological maturity, realisation times, CO_2 emissions avoided, visual impact, acoustic impact, impact on ecosystems, social acceptability	Wind energy installations
Madlener and Stagl (2005)	Austria	PROMETHEE	Biophysical dimension: resource inputs needed for production (land resources, water, material requirements, indirect energy requirements), potential environmental consequences (impact on natural biota, habitats and wildlife, environmental risks, visual intrusion, impact on microclimate, impact on soil productivity, impact of resettlements), potential consequences of energy conversion and use (air pollution, organic emissions, generation of solid wastes, water pollution, pressure on land and water resources and other hazards), and socio-economic impacts (employment, occupational hazards, noise, impact on local poverty, household income disparity, democratic control over markets, safety of power supply, impact on balance of trade, long-term economic viability, local net value added, economic risk to ratepayers, impact on flexibility of supply)	Hydro, wood, wind, and PV
Gamboa and Munda (2007)	Catalonia, Spain	NAIADE	Land owner's income, distribution of income, income of municipalities, number of jobs, visual impact, forest loss, noise nuisance, avoided CO_2 emissions, and installed capacity	Wind energy installations

(continued)

Table 8.1 (continued)

Authors and year	Application	MCDA method	Criteria	Alternatives
Burton and Hubacek (2007)	Municipality in Yorkshire, UK	MACBETH	Capital cost, operation and maintenance, generation capacity, lifespan, carbon emissions, noise, natural environment and social consequences	Solar PV, micro-hydro, micro-wind, biomass, large scale wind, landfill gas, large scale hydro, energy from waste
Diakoulaki and Karangelis (2007)	Greece	Promethee	Investment cost, production cost, guaranteed energy (10), available power during peak load, security of supply, CO_2 increase, SO_2 emissions, and NOx emissions	Energy strategies
Chatzimouratidis and Pilavachi (2008)		AHP	Quality of life (accident fatalities, NMVOCs, CO_2-eq, NOx, SO_2, PM and land required) and socio-economic aspects (job creation, compensation rates, social acceptance)	Coal, oil, natural gas turbine, natural gas combined cycle, nuclear, hydro, wind, photovoltaic, biomass, and geothermal

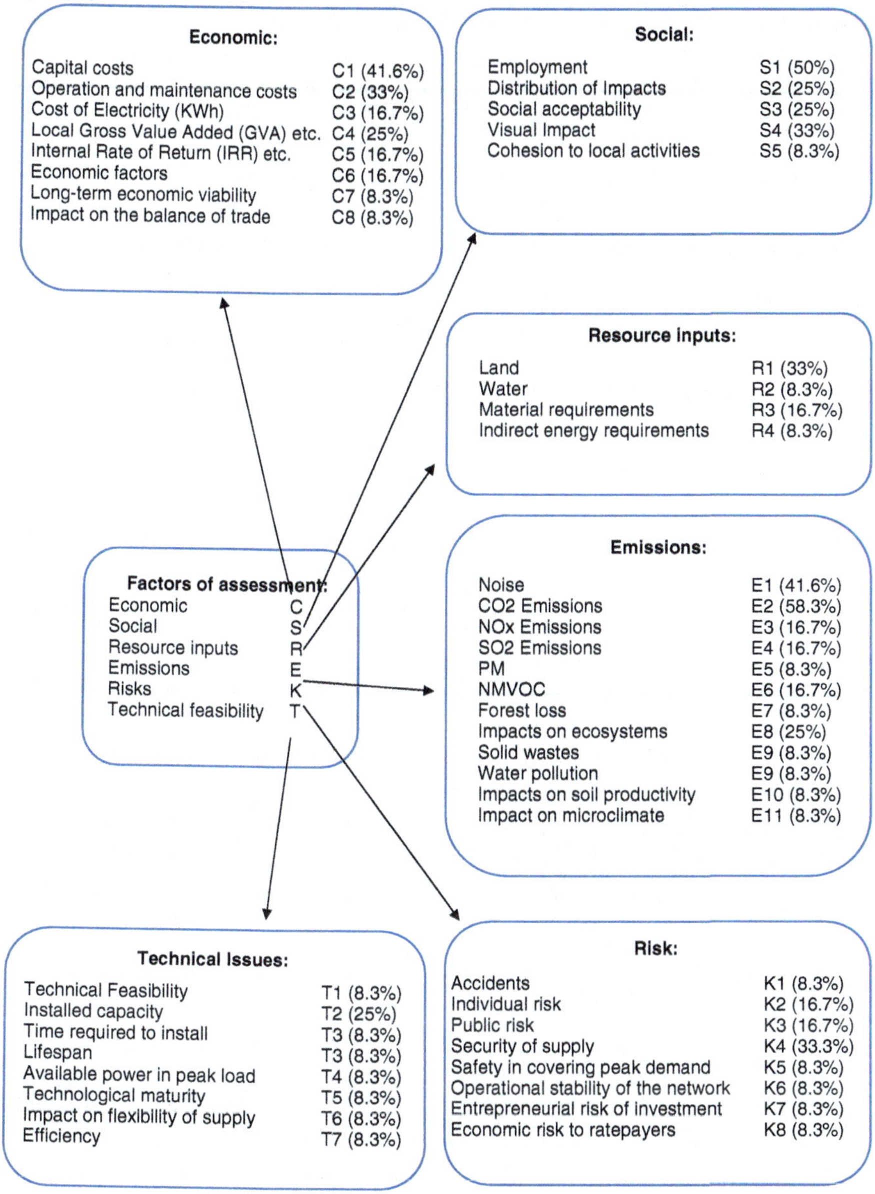

Fig. 8.1 Factors of sustainable energy systems analysis

Percentages given next to each criterion denote the frequency of occurrence of each criterion in the body of research. For example, in the Resource Input section, 33% of the studies took into account "Land" as a criterion, 8.3% assessed the use of "Water", 16.7% took into account "Material Requirements", and 8.3% – "Indirect Energy Requirements". The most popular criteria have been: "Capital costs" and "Operation and Maintenance Costs" in the economic set; "Employment" and "Visual Impact" in the social set; "CO_2 emissions" and "Noise" – in the emissions set; "Installed Capacity" in the technical set; "Land" in the resource inputs set and "Security of Supply" in the risks set. The chart represents the result of substantial analytical work and gives a very good overview of how sustainable energy issues were addressed in the past. It should be noted perhaps that the frequency of occurrence of a criterion in the literature does not mean that it is the only necessary factor to be taken into account, quite the contrary: it makes sense to study all the criteria identified and form a selection according to the goals of a particular project.

It would be interesting to compare existing tools used for energy planning and modelling, such as MARKAL with the taxonomy of criteria presented in Fig. 8.1. It will be clear that the absolute majority of studies do not cover even half of all the important aspects of the problem which need to be taken into account.

Decision Support Systems

Decision support for sustainable energy systems is a complex task and usually requires the application of sophisticated software. Some of the major software packages which might be used for sustainable energy analysis are compared in Table 8.2. Of all the tools, the following were selected: DREAM (Open University), LEAP, RetScreen, GEMIS, MDM E3, Homer, MARKAL, MESSAGE. All the packages are compared using the same framework and the analysis is focused on the following parameters: main purpose; integratedness; main instrument being used; special capabilities; static/dynamic character; timescale of the possible model; geographical scope; non-energy effects taken into account; sectors of the economy; cost of the model, how widely it is used and who developed it.

It becomes apparent that, in order to chose the most relevant model for the task, one needs to travel across the table and check whether the time scale offered, the geographical scope, number and types of technologies being taken into account, the detailed description of economic links, the type of instrument used (systems dynamics, econometrics, input–output analysis, or optimization) and data availability to fill the model fit the purpose of the assessment and the amount of information available.

Often, no single tool is capable of addressing all the research needs of a particular project. Sometimes, new modelling effort will be required; we live in a very complex world, so why should we address it with oversimplified methods of enquiry?

Renewable Energy in the UK

Let us turn to the current state of affairs in renewable energy which, after the recent nuclear tragedy in Japan, will become more politically interesting. It is clear that at the present moment renewable energy alone cannot satisfy all the world's energy requirements and nuclear energy plays a very important role; however it is clear that the future should depend on the new generation of more environmentally friendly energy-generation technologies.

The World Wind Energy Report published by the World Wind Energy Association announced that in 2008 the total world installed capacity of wind has reached 121′188 MW, of which almost a quarter, 27′261 MW were added in 2008. All wind turbines installed by the end of 2008 worldwide generate 260 TWh per annum, equalling more than 1.5% of global electricity consumption. Wind energy became a global job creator and generated 440,000 jobs worldwide. In 2008, the total turnover of the wind energy sector has reached a mark of 40 billion euro. USA and Germany took the lead in share of world installed capacity: 25,170 and 23,903 MW respectively. UK, with only 3,200 MW, follows in 8th place after Spain, China, India, Italy and France. The UK installed capacity in 2008 was equal to one thirds of the wind capacity in India and one fifth of that of Spain. The share of renewable energy in final energy consumption in the UK is currently 0.53%.

Figure 8.2 represents UK final energy consumption in 2007. We can note that a very large share of the mix is occupied by petroleum products (48%), and the second largest category is natural gas (31%) with electricity occupying the third place at almost 18%. It should be noted that a major proportion of petroleum products goes into the manufacturing of car and aviation fuel, and considerable amounts of gas and electricity are used in homes.

If we explore the dynamics of renewable energy generation in the UK (Fig. 8.3) it will be clear that the UK managed to more than double the amount of energy generated from renewables between 2003 and 2009, with the most rapidly expanding sectors being onshore wind, off-shore wind, and landfill gas, given that large scale hydro had a stable presence in the mix.

The UK Government is currently trying to establish its long-term strategy devoted to the radical reduction of CO_2 emissions, primarily caused by the generation of energy and transport activities. As was shown earlier, there are substantially different possible strategies for achieving this, namely, using substantial proportions of nuclear energy, natural gas and off-shore wind or a balanced mix of smaller-scale renewables including on-shore wind, hydro, geothermal, solar and other sources. Characteristically, these paths could have surprisingly similar CO_2 generation trends.

The MARKAL Model

To support national-scale decision making on CO_2 emissions mitigation a model called MARKAL is very often used. This model is an optimization tool developed by/for the IEA to assess various strategies of development of the energy system. It is a technological choice model which operates in terms of costs and emissions associated with different technologies and it is usually run in a cost-minimization setting. Historically, the MARKAL model, developed at the International Energy Agency (Rath-Nagel and Stocks 1982) in the aftermath of the 1970s energy crisis,

Table 8.2 Sustainable energy assessment software

	DREAM	LEAP	RetScreen	GEMIS
Main purpose	Dynamic regional energy analysis	Long range energy alternatives planning system	Clean energy project analysis software	Global emission model for integrated systems
Integrated-ness	Integrated	Integrated	Series of technology oriented models	Integrated
Main instrument	Systems dynamics modelling	Scenario analysis	Project analysis	Lifecycle analysis; scenario/process analysis
Special capabilities	Analysis and visual representation of key CO_2 emission and abatement policy scenarios	– Energy policy analysis; – Environmental Policy Analysis; – Biomass and Land-Use Assessment; – Investment Project Analysis; – Integrated Energy Planning; – Full Fuel Cycle Analysis; – Energy balances, – Lifecycle diagrams, – Costs and benefits, fuel transport	1. Clean Energy Project Analysis; 2. GHG Emission analysis; 3. Financial Risk Analysis; Sensitivity analysis	1. Lifecycle computations for a variety of emissions, estimation of resource use; 2. Costs calculation; 3. Aggregation of emissions into CO_2 equivalents, calculation of external costs; 4. emission standards, 5. energy carriers
Static-dynamic	Dynamic	Simple dynamics, growth rates, Time-series wizard	Dynamic	Dynamic

Cambridge MDM E3 model	Homer	Markal	MESSAGE
Multisectoral dynamic model of the UK economy	Micropower optimization model	Energy supply with constraints	Optimisation of energy supply systems
Integrated	Integrated	Integrated	Integrated
Econometric input–output modelling	Optimisation and sensitivity analysis	Linear programming, dynamic programming	Dynamic programming
Analysis of effects of environmental policy, changes in energy prices, development in particular sectors of the economy on the rest of the economy and the environment; Energy–Economy–Environment forecasts; Capability to take long-term development into account	Simulation: HOMER simulates the operation of a system by making energy balance calculations for each of the 8,760 h in a year. Optimization After simulating all of the possible system configurations, HOMER displays a list of configurations, sorted by	1. To identify least-cost energy systems; 2. To perform prospective analysis of long-term energy balances under different scenarios; 3. to evaluate new technologies and priorities for R&D; 4. to evaluate the effects of regulations, taxes, and subsidies;	Detailed description of energy end uses and (renewable) energy technologies. MESSAGE III evaluates energy systems costs and capital requirements for energy planning (scenario development) and CO_2 tax impacts on energy mix
Dynamic	Dynamic	Dynamic	Dynamic

Table 8.2 (continued)

	DREAM	LEAP	RetScreen	GEMIS
Time scale	Monthly periods	Medium to long-term, annual time steps	25 years	1995–2030
Geographic scope	Regional/National	Global/Regional/ National/Local	Local with any location globally	Regional
Non-energy effects	CO_2 emissions	Cement process emissions: CO_2 Non-biogenic landfill emissions: methane; CO_2 sequestration, CO_2 Non-biogenic	Tonnes of CO_2 per year; CH_4 and N_2O emissions converted to CO_2 equivalent	CO_2 equivalents, SO_2 equiva-lents, and Tropospheric ozone precursor potential (TOPP), and by a calculation of external costs; Employment effects; Resources Use: CEC, CER, and CMR; Land Use – area affected by processes; Costs – internal and external costs

Cambridge MDM E3 model	Homer	Markal	MESSAGE
1960–2010	8760 1 h periods per year	Medium, long term (40–50 years): minimum 1 year; maximum 9 periods; 9 periods of 5 years is most common	Short, medium, long term
National (UK-MDM-E3), European E3ME, Global E3MG (being developed)	Local/Regional	Local, regional, national scales possible	Global (11 Regions)
Employment, Impacts on related sectors and the economy as a whole, emissions	Different designs of the renewable energy system; include possibility to sell to the grid.	CO_2 emissions, costs, consumer surplus	Resource extraction analysis;-import/export of energy;-energy conversion analysis;-energy transport and distribution analysis;- final energy utilisation by consumers analysis;-environmental protection policy;-investment policy;-opportunity costs

Table 8.2 (continued)

	DREAM	LEAP	RetScreen	GEMIS
Sectors of the economy	Domestic Sector (space heating, water heating, cooking, lighting and appliances); Services Sector (Offices, Pubs and Clubs, Residential (Hotels etc.), Commercial Services (Banks etc.), Government, Shops, Distribution, Catering, Defence, Education, Health,	Household (urban and rural); Industry (Iron and Steel, Pulp and Paper etc.); Transport; Commercial	Renewable energy sectors (project specific)	NACE Nomenclature (99 sectors)
Cost	Free to OU	Free for an evaluation version	Free	Free
How widely used?	Milton Keynes, Barcelona	100 government agencies, utilities and research institutes in over 30 countries (Costa Rica, Senegal, Philippines, USA, Zimbabwe etc.)	Project management and regional authorities, Canada, USA, Germany, UK, Denmark, Japan, France, China etc.	Universities, companies and local authorities (mostly Germany)
Developer	Dr Godfrey Boyle, The Open University	The Boston Center of the Stockholm Environment Institute (SEI Boston)	Natural Resources Canada	Öko-Institut e. V. (Institute of Applied Ecology, Germany)

Cambridge MDM E3 model	Homer	Markal	MESSAGE
41 industries	Renewable energy sectors (project specific)	Energy sector only	Energy sector
Not for sale	Free	Expensive	No data
UK government	Almost 9000 users in 178 countries	Most IEA member states, Indonesia, Brazil, Tunisia, local communities, regional utilities, etc.	IPCC
Cambridge Econometrics	National Renewable Energy Laboratory, USA	International Energy Agency (IEA)/ETSAP	International Institute for Applied System Analysis (IIASA), Austria

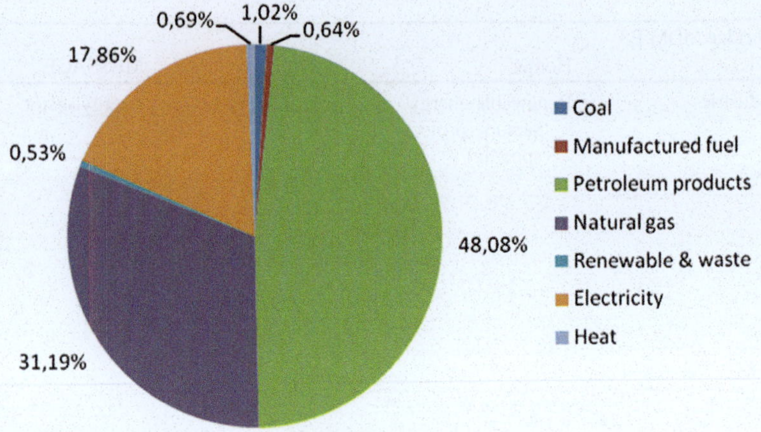

Fig. 8.2 Final energy consumption, UK, 2007

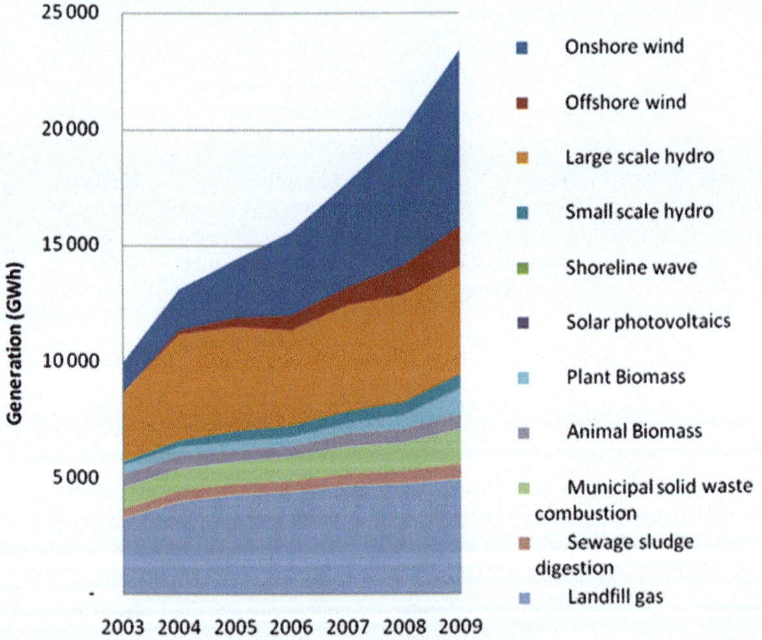

Fig. 8.3 UK energy generated from renewables GWh

was widely used to address the needs of strategic policy formulation related to the changing energy mix. It is interesting to note that in the original paper, the MARKAL model has been presented as a multicriteria optimization tool, with a focus on analysing efficiency frontiers or the boundaries of non-dominated solutions. Over the years, the attention of the energy planning community diverged from the initial formulation of the energy model and it was simplified to single criterion optimization, which, unfortunately, neglects very important aspects of the problem.

The MARKAL model with its extensions is currently used in 79 institutions in 38 countries: it has been applied in Australia (Naughten 2002), Department of Energy's Office of Policy used MARKAL as the primary tool to analyse the impact of the Kyoto Protocol on the U.S. energy system USA (Morris et al. 2002), the UK (Strachan et al. 2007), the province of Ontario, Canada (Berger et al. 1990), China (Chen 2005), the Netherlands (Gielen 1995), Latvia (Shipkovs et al. 1999), Estonia (Agabus et al. 2007), Switzerland (Schulz et al. 2008), Vietnam (Nguyen 2007). Applications focused on particular technologies or policy instruments deal with, among others, green certificate market in the Nordic countries (Unger and Ahlgren 2005) and in Italy (Contaldi et al. 2007), photovoltaics (Endo and Ichinohe 2006), vehicle mix in the passenger car sector in Japan (Ichinohe and Endo 2006).

Just as, in the financial sector, the use of the same asset pricing model might lead to emerging behaviour for the system as a whole which could lead to a crash, the energy system is no exception. If, following a recommendation from the International Energy Agency, most OECD countries use a cost-minimising MARKAL model to design their energy strategies, it is not surprising that renewables have not reached a share close to 30–50% in most European countries. There seems to be a serious problem in using a single criterion optimization framework for forward-looking-innovation-led strategic planning. It is absolutely clear that the original intention of using multi-criteria optimization for energy systems modelling was more appropriate for the type of system the energy system is, given the multiple dimensions of its performance, depicted in Fig. 8.1.

Tools as alternative to optimization or multicriteria decision aid have also been used; the analysis of the energy system has been pioneered in the 1970s by John Proops, who employed the technique of input–output analysis (Gay and Proops 1993, Proops 1977, 1984, Proops and Speck 1996. A Cambridge group of econometric input–output modellers, which grew out of the Cambridge Growth project chaired by Richard Stone, has produced one of the most detailed energy–environment–economy models of the UK economy, entitled MDM-E3 (Barker 1981, Barker et al. 1980a, 2007a, b, c).

It seems that, given the current development of mathematical methods, an area of multicriteria optimization of national sustainable energy systems becomes very promising. Chapter 11 deals with a similar issue in the context of sustainable waste management. The interested reader will note that mathematically those two problems are almost identical. It would make perfect sense to explore the so called "efficient sets" of non-dominated solutions and create new and useful tools for comparing the elements of those sets. Also it would be highly desirable to undertake full comparative analysis of various energy technologies either taken separately or in various mixes using the taxonomy of criteria outlined in Fig. 8.1. In this book we cannot provide all the answers, something must be left unexplained and it is up to the interested reader to explore the potentially challenging but highly stimulating problems at which this book hints.

References

Afgan NH and Carvalho MG (2002) Multi-criteria Assessment of New and Renewable Energy Power Plants *Energy* 27.8 739-755

Agabus H, Landsberg M and Tammoja H (2007) Reduction of CO_2 Emissions in Estonia During 2000–2030 *Oil Shale* 24 209-224

Barker T (1981) Projecting Economic Structure with a Large-Scale Econometric Model *Futures* 13.6 458-467

Barker T et al (1980a) The Cambridge Multisectoral Dynamic Model: An Instrument for National Economic Policy Analysis *Journal of Policy Modeling* 2.3 319-344

Barker T, Ekins P and Foxon T (2007a) Macroeconomic Effects of Efficiency Policies for Energy-Intensive Industries: The Case of the UK Climate Change Agreements, 2000–2010 *Energy Economics* 29.4 760-778

Barker T, Ekins P and Foxon T (2007b) The Macro-Economic Rebound Effect and the UK Economy *Energy Policy* 35.10 4935-4946

Barker T, Junankar S, et al. (2007c) Carbon Leakage from Unilateral Environmental Tax Reforms in Europe, 1995â"2005 *Energy Policy* 35.12 6281-6292

Berger C et al (1990) Modelling Energy Use in the Mineral Processing Industries of Ontario with MARKAL-Ontario *Energy* 15.9 741-758

Burton J and Hubacek K (2007) Is Small Beautiful? A Multicriteria Assessment of Small-Scale Energy Technology Applications in Local Governments *Energy Policy* 35.12 6402-6412

Cavallaro F and Ciraolo L (2005) A Multicriteria Approach to Evaluate Wind Energy Plants on an Italian Island *Energy Policy* 33.2 235-244

Chatzimouratidis AI and Pilavachi PA (2008) Sensitivity Analysis of the Evaluation of Power Plants Impact on the Living Standard Using the Analytic Hierarchy Process *Energy Conversion and Management* 49.12 3599-3611

Chen W (2005) The Costs of Mitigating Carbon Emissions in China: Findings from China MARKAL-MACRO Modeling *Energy Policy* 33.7 885-896

Contaldi M, Gracceva F and Tosato G (2007) Evaluation of Green-Certificates Policies Using the MARKAL-MACRO-Italy Model *Energy Policy* 35.2 797-808

Diakoulaki D and Karangelis F (2007) Multi-criteria Decision Analysis and Cost-Benefit Analysis of Alternative Scenarios for the Power Generation Sector in Greece *Renewable and Sustainable Energy Reviews* 11.4 716-727

Endo E and Ichinohe M (2006) Analysis on Market Deployment of Photovoltaics in Japan by Using Energy System Model MARKAL *Solar Energy Materials and Solar Cells* 90.18–19 3061-3067

Gamboa G and Munda G (2007) The Problem of Windfarm Location: A Social Multi-criteria Evaluation Framework *Energy Policy* 35.3 1564-1583

Gay PW and Proops John LR (1993) Carbon Dioxide Production by the UK Economy: An Input-Output Assessment *Applied Energy* 44.2 113-130

Georgopoulou E, Lalas D and Papagiannakis L (1997) A Multicriteria Decision Aid Approach for Energy Planning Problems: The Case of Renewable Energy Option *European Journal of Operational Research* 103.1 38-54

Gielen D (1995) Toward Integrated Energy and Materials Policies?: A Case Study on CO_2 Reduction in the Netherlands *Energy Policy* 23.12 1049-1062

Haralambopoulos DA and Polatidis H (2003) Renewable Energy Projects: Structuring a Multi-criteria Group Decision-Making Framework *Renewable Energy* 28.6 961-973

Ichinohe M and Endo E (2006) Analysis of the Vehicle Mix in the Passenger-Car Sector in Japan for CO_2 Emissions Reduction by a MARKAL Model *Applied Energy* 83.10 1047-1061

Madlener R and Stagl S (2005) Sustainability-Guided Promotion of Renewable Electricity Generation *Ecological Economics* 53.2 147-167

Mavrotas G, Diakoulaki D and Capros P (2003) Combined MCDA-IP Approach for Project Selection in the Electricity Market *Annals of Operations Research* 120.1–4 159-170

Morris SC, Goldstein GA and Fthenakis VM (2002) NEMS and MARKAL-MACRO Models for Energy-Environmental-Economic Analysis: A Comparison of the Electricity and Carbon Reduction Projections *Environmental Modeling and Assessment* 7.3 207-216

Naughten B (2002) Some MARKAL capabilities inAustralian energy policy analysis and modeling the mandated target for renewable electricity in Australian MARKAL Available at http://adl.brs.gov.au/data/warehouse/pe_abarebrs99000834/PC12388.pdf

Nguyen KQ (2007) Impacts of Wind Power Generation and CO_2 Emission Constraints on the Future Choice of Fuels and Technologies in the Power Sector of Vietnam *Energy Policy* 35.4 2305-2312

Noble BF (2004) A Multicriteria Analysis of Canadian Electricity Supply Futures *The Canadian Geographer/Le GeÂ´ographe canadien* 48.1 11-28

Proops JLR (1977) Input-Output Analysis and Energy Intensities: A Comparison of Some Methodologies *Applied Mathematical Modelling* 1.4 181-186

Proops John LR (1984) Modelling the Energy-Output Ratio *Energy Economics* 6.1 47-51

Proops John LR and Speck S (1996) Comparison of Neoclassical and Neo-Austrian Growth Models *Structural Change and Economic Dynamics* 7.2 173-192

Rath-Nagel S and Stocks K (1982) Energy modelling for technology assessment: The MARKAL approach. *Omega,* 10.5 493-505

Schulz TF et al (2008) Intermediate Steps Towards the 2000 W Society in Switzerland: An Energy-Economic Scenario Analysis *Energy Policy* 36.4 1303-1317

Shipkovs P, Kashkarova G and Shipkovs M (1999) Renewable Energy Utilization in Latvia *Renewable Energy* 16.1–4 1241-1244

Siskos J and Hubert P (1983) Multi-criteria Analysis of the Impacts of Energy Alternatives: A Survey and a New Comparative Approach *European Journal of Operational Research* 13.3 278-299

Strachan N, Kannan R and Pye S (2007) Final Report on DTI-DEFRA Scenarios and Sensitivities Using the UK MARKAL and MARKAL MACRO Energy System Models

Unger T and Ahlgren EO (2005) Impacts of a Common Green Certificate Market on Electricity and CO_2-Emission Markets in the Nordic Countries *Energy Policy* 33.16 2152-2163

Chapter 9
Biodiversity Loss: New Methods for Evaluating Ecosystems

Abstract This chapter summarises research undertaken to develop a methodology for a multi-criteria assessment of biodiversity which takes into account a multitude of criteria and stakeholder perspectives. The proposed methodology will be of particular value for developing countries, where conflicts of interest regarding ecosystems and biodiversity are numerous and often involve businesses, government, local residents, and other stakeholders. The chapter reviews the state of the art in the field of multi-criteria methods and assessment of ecosystems and biodiversity. It presents the results of analytical work undertaken on the basis of interviews carried out in the Provence–Alpes–Côte d'Azur (PACA) region of France, focusing on biodiversity in the Réserve Naturelle Coussouls de Crau.

The chapter addresses three main issues: selection of the multi-criteria assessment method, selection of the assessment criteria, and a comparison of stakeholder interests in the context of biodiversity analysis. The identification of potential decision criteria was based on a survey of key stakeholders, namely the Management of the Réserve Naturelle Coussouls de Crau; the Muséum National d'Histoire Naturelle, a national biodiversity research institution; the Laissez-faire Association, protecting the interests of the agricultural community; CDC Biodiversité (a branch of Caisse des Dépôts), a group carrying out long-term investments in the public interest; and the Direction regionale de l'environnement Provence–Alpes–Côte d'Azur (DIREN-PACA). Based on these interviews, 14 ecological, nine economic, and 12 social criteria were identified. Further analysis revealed very few points of overlap among the interests of the stakeholders, which complicates the case for consensus building.

Not accepting the idea that the value of ecosystems and biodiversity can be expressed in monetary terms, the author suggests an alternative, more inclusive approach, focusing on multiple social, economic, and ecological dimensions of ecosystem value, and illustrates the existence of divergent interests among stakeholders. This experience would be particularly useful in situations where local communities have to defend their right to a clean environment and preserve important virgin ecosystems for future generations.

S.E. Shmelev, *Ecological Economics: Sustainability in Practice*,
DOI 10.1007/978-94-007-1972-9_9,

Keywords Ecosystems • Biodiversity • Multi-criteria decision aid (MCDA) • Stakeholders • Criteria • Interests

Ecosystems

Ecosystems can be seen as a scarce common-pool resource with a multitude of characteristics, seen by potential users from differing perspectives (Adams et al. 2003, Lant et al. 2008, Ostrom 2008, Ostrom et al. 1999). According to the Millennium Ecosystem Assessment, "over the past 50 years, humans have changed ecosystems more rapidly and extensively than in any comparable period of time in human history, largely to meet rapidly growing demands for food, fresh water, timber, fibre, and fuel. This has resulted in a substantial and largely irreversible loss in the diversity of life on Earth" (Millenium Ecosystem Assessment 2004). There is a need to facilitate decisions on the future of ecosystems and to have clear methodologies for classifying ecosystems into valuable ecosystems which should be protected, parts that should be restored and parts that might be developed. The instrument of multidimensional integrated assessment might be a good tool to manage the complex resources under study and help to reduce the fragmentation of ecosystems and improve their quality. The key question when setting up a mitigation banking system becomes how to classify or value ecosystems, what might be the unit of value (if anything) and how to find an area of equal value to compensate within the mitigation banking mechanism?

Ecosystems are multifunctional, complex systems, described by a multitude of characteristics from the point of view of multiple criteria (Fig. 9.2). How to compare objects with multiple characteristics has been the focus of area of Multi-Criteria Decision Aid (MCDA). Methods of multi-criteria analysis have been developed to address the problem of incommensurable values. These methods seek to account for the social, economic and environmental dimensions of decisions. This paper presents a review of methods and applications of multi-criteria analysis in the context of ecosystems and biodiversity assessment, and a selection of the most appropriate tool among the MCDA methods and it identifies a set of criteria relevant to the case study in southern France. A discussion on the diversity of interests and ways of mitigating conflict will follow.

Recently, at regional, corporate and local level, decisions regarding the use of natural resources, investments and other forward looking strategies have been guided by monetary methods, notably cost-benefit analysis. Following Kapp (1970), O'Neil (1997), Foster (1997), and Martinez-Alier et al. (1998) showed the role which incommensurability of values plays in decision making problems, illustrated the inherent limitations of cost-benefit analysis, and identified multicriteria methods as viable alternatives. Sustainability problems usually imply relatively low levels of substitutability among criteria to be satisfied, given the urgency and complexity of the problems we are facing: loss of biodiversity, climate change, deterioration of public health, and poverty. We need to understand multiple dimensions of the decisions that are taken today and the links between these dimensions (Shmelev and Shmeleva 2009). In making decisions, it is necessary to assess likely multiple consequences of these decisions in the future and to work on innovative strategies which would satisfy multiple criteria to the best possible extent. New multicriteria methods, when applied at the local, regional and corporate level would allow stimulation of the shift in the development pattern toward sustainability.

Provence Case Study

The Nature Reserve of Crau is situated in the South of France, southeast of Arles. The region is surrounded from the South by the Mediterranean Sea, from the East by Étang de Berre, and from the West by the river Rhone. The region is well described in the paper by Buisson (Buisson and Dutoit 2006). Figure 9.1 depicts the physical geography of the region of Crau. The Natural Reserve of Crau is embedded within a complex network of environmentally sensitive and protected territories. From the West it is surrounded by a large RAMSAR site. The PACA region is covered by a few larger and smaller National and Regional Nature Reserves. The region is also neatly covered by the network of marine and land based Zones Naturelles d'Intérêt Ecologique Faunistique et Floristique (ZNIEFF). In addition, parts of the Crau region are covered by the system of Reserves de Biosphere. The Crau region does not have any Reserves Biologiques, or Arretes de Protection de Biotope, neither does it have any Parcs Nationaux. However, it is adjoined by the Parc Naturel Regional of the Camargue. The system of Natura 2000 territories,[1] largely different from the types mentioned above is extremely diverse and covers a considerable proportion of the region

If we look more closely at the Crau region, the patchiness and multiple designations of the same little territories becomes apparent. It is often the case that a small piece of land is designated simultaneously as a Reserve Naturelle National and Parc National Regional or, a Parc Naturelle Regional may be part of a Ramsar site. The full structure of the multiplicity of designations in the Crau Region is depicted in Table 9.1. It should be noted that the value of each individual site to be assessed with the help of MCDA would become larger if it fell into several designation categories.

[1]Natura 2000 is an EU-wide network of Special Areas of Conservation (SAC) and Special Protection Areas (SPA). Special Protection Areas (SPAs) are created applying an E.U. directive requiring the protection of wild birds (79/409/EEC, 1979). The Department of Ecology of each country designates these areas and potential management is planed locally (France: 103 SPA; 8,000 km^2). The SPA Crau seche was designated in 1990 and covers 11,816 ha. Special Conservation Areas (SCAs) are designated applying annex II (animal and plant species of community interest) of the E.U. directive requiring the protection of natural habitats (92/43/EEC, 1992). Annex I plans for the establishment of a consistent network of SCAs within which SPAs are automatically integrated: NATURA 2000. The SCA Crau centrale – Crau se`che was designated in 1996 and covers 31,458 ha. (Buisson and Dutoit 2006).

Fig. 9.1 Nature reserve of Crau

The general problem decision makers are facing in the region is the comparison of 60–70 sites within or adjacent to a Nature Reserve and the estimation of which sites should be incorporated into the Nature Reserve and which given to developers for such projects as a gas pipeline.

In cases when there is only one criterion and an infinite number of alternatives, single criterion optimization is usually the most appropriate tool; when both the number of alternatives is infinite and the number of criteria is more than one, an apparatus of multicriteria optimization may be applied. In the situation, where the number of criteria is more than one and the number of objects to be compared is finite, MCDA proves to be a viable tool for the development of a robust scientific assessment methodology, which can be replicated. Alternative approaches, namely citizens' juries, can be considered, but are not practical because of the relatively high number of alternatives (60–70), and also given the existing time constraints

Table 9.1 Multiple designations in the Crau region

	National Nature Reserves	Regional Nature Reserves	Nature Parks (PN)	Regional Nature Parks (PNR)	ZNIEFF Sea	ZNIEFF Land	Biosphere Reserves (RBS)	RAMSAR Sites	Natura 2000 Birds	Natura 2000 Habitats
National Nature Reserves (RNN)				X		X	X	X	X	X
Regional Nature Reserves (RNR)				X		X	X	X	X	X
Nature Parks (PN)										
Regional Nature Parks (PNR)	X	X			X	X	X	X	X	X
ZNIEFF Sea				X			X	X	X	X
ZNIEFF Land	X	X		X			X	X	X	X
Biosphere Reserves (RBS)	X	X		X	X	X		X	X	X
RAMSAR Sites	X	X		X	X	X	X		X	X
Natura 2000 Birds	X	X		X	X	X	X	X		X
Natura 2000 Habitats	X	X		X	X	X	X	X	X	

and budget limitations; however the Delphi method of using expertise and interacting with stakeholders will be applied in this study. The chapter will focus on three main issues: identification of the assessment criteria, selection of the multi-criteria assessment method and the comparison of stakeholder interests in the context of ecosystem and biodiversity assessment.

Integrating Socio-Economic Information in Conservation Planning: A Multi-Criteria Framework

Multi-criteria evaluation of biodiversity for the purposes of ecological compensation and mitigation banking[2] present a methodological as well as a practical challenge. Multi-criteria decision tools allow simultaneous consideration of a wide range of criteria, representing different dimensions of sustainability. These may include: poverty; governance; health; education; demographics; natural hazards; atmosphere; land; oceans, seas and coasts; freshwater, biodiversity, economic development, global economic partnership, consumption and production patterns (United Nations 2007) or the social, environmental and economic dimensions of sustainability in the previous edition of the United Nations Guidelines. The latest edition of the UN Guidelines on indicators of sustainable development (United Nations 2007) emphasises the link between different dimensions of sustainability: e.g. the indicator "Percentage of trees damaged by defoliation" is related to the key thematic area "Land" as well as Biodiversity, and Consumption and Production Patterns. "Fragmentation of habitat" is related to the key thematic area "Biodiversity" as well as to Governance, Land, and Consumption and Production Patterns.

In the ecological or more broadly, natural science domain, recent research in earth systems science and complexity by V. G. Gorshkov et al. (2000), J. Lovelock (1992), S. Harding (2004) shows how the complexity of ecosystems and the ecological web and the biosphere in general can determine climatic stability and the resilience of the surrounding region or the global system. Recent research by Robert Costanza (2008) contributes to the debate on the evaluation of a multitude of ecosystem services. Thus modern science reveals the increasing importance of cross-disciplinary feedback loops.

Regan et al. (2007) present a coherent set of environmental criteria for evaluating biodiversity. Alexander Moffet (2006) offers an extensive overview of existing applications of multi-criteria methods to the problem of biodiversity evaluation. It is interesting to note that the majority of studies reviewed in this paper have been carried out with the help of the MAVT, AHP or goal programming methods.

[2]Mitigation banking is a tool which emerged in the USA in the 1970s in order to diminish the loss of wetland caused by development projects as required by the federal Clean Water Act of 1972. The main function of a mitigation bank is to compensate for adverse impacts on natural resources by providing for the conservation of a similar resource in another location.

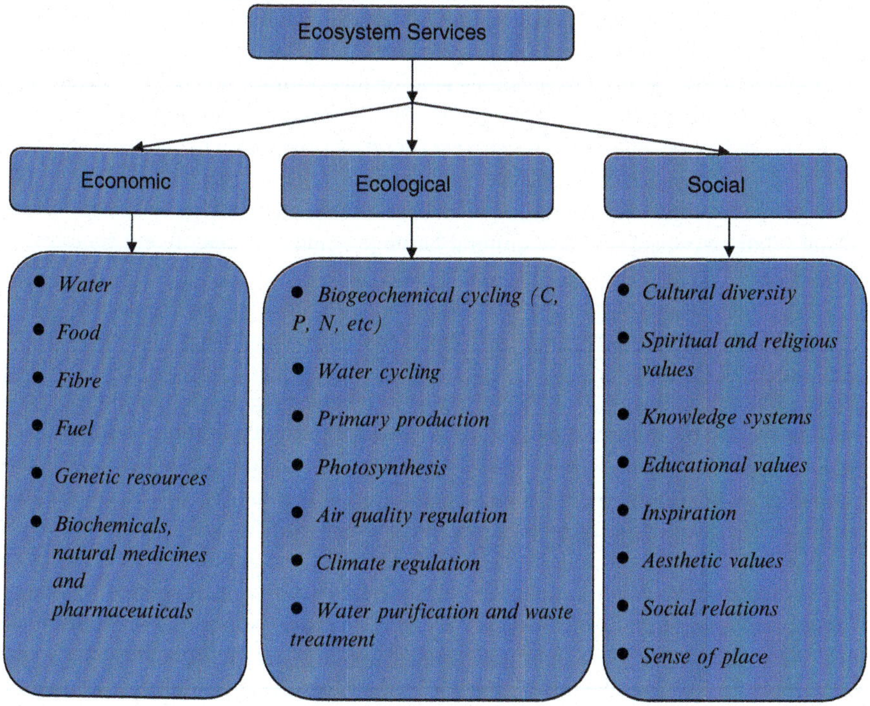

Fig. 9.2 Classification of ecosystem services (Adapted from the Millennium Ecosystem Assessment 2004)

It should also be noted that the use of social criteria has been particularly rare in any multi-criteria evaluation of biodiversity.

In the economic domain, in the spirit of Pearce and Moran (1994) and Costanza et al. (1997), increasing efforts have been devoted to attempts to estimate, in economic terms, the value of ecosystem services and biodiversity; this had two major aims: to attract public attention to the problem of ecosystem services (which it was successful in doing) and to provide the basis for decision making (at which it was less successful). Methods illustrating the economic value of biodiversity have been reviewed by Jeroen van den Bergh and Paola Nunes (Nunes and van den Bergh 2001), who concluded that the empirical literature fails to apply economic valuation to the entire range of biodiversity benefits. Therefore, available economic valuation estimates should generally be regarded as providing an incomplete perspective on, and at best a lower boundary to, the unknown value of biodiversity changes. The attempt to assess the value of ecosystem services and biodiversity using a single criterion of money is clearly a simplification in a degree higher than world ecosystems can bear. In our case, incommensurability of certain aspects of value, which is essentially, a multidimensional concept, plays a crucial role. This means that the value of ecosystems and biodiversity should be considered using multicriteria methods, which correspond nicely to the multiple dimensions of ecosystem value (see Fig. 9.2).

In the social domain, several studies have identified the following socio-political criteria for biodiversity assessments: economic cost, recreational value, human population, future economic value, scenic beauty, cultural heritage, educational value (Alexander Moffett 2006). The Millennium Ecosystem Assessment (Millenium Ecosystem Assessment 2004) describes the following cultural services provided by ecosystems: cultural diversity, spiritual and religious values, knowledge systems (traditional and formal), educational values, inspiration, aesthetic values, social relations, sense of place, cultural heritage values, recreation and ecotourism (Fig. 9.2).

The social and cultural aspects of biodiversity conservation, ecosystem health and landscape quality have been addressed in a veritable cornucopia of literature (Gehlbach 1975, Inhaber 1977, Peterson 1974, Ploeg and Vlijm 1978, Tubbs and Blackwood 1971, Wright 1977, Zube et al. 1982). The diversity of the landscape has been proven to be an important feature in providing visual comfort to humans: in a series of seminal papers Ulrich (1979, 1986) showed how important trees and vegetation in general are for the health and wellbeing of an individual. Modern research in the public evaluation of landscapes (Nijnik et al. 2008, Ode et al. 2009, Tveit et al. 2006) shows how a multitude of approaches can be applied to the analysis of social preferences which stakeholders have towards different scenarios for biodiversity and natural ecosystems. Diversity of the landscape is often reported as an important factor for the visual satisfaction humans derive from observing and experiencing scenery. Fragmentation has been identified by Taylor (2002) as an important issue in the field of landscape research and planning, which should be looked at from different points of view: ecological, socio-cultural, anthropological. Lausch and Herzog (2002), and Li and Wu (2004) discuss a range of landscape metrics used for the study of regional environmental change, data availability and analytical procedures for landscape research. Other integrative attempts to evaluate the quality of landscapes include: (Antrop and Eetvelde 2000, Arriaza et al. 2004, Coeterier 2002, Dramstad et al. 2006, De Groot and Van Den Born 2003, de la Fuente de Val et al. 2006). The environmental psychology school has also made interesting contributions to this field (Hagerhall 2001, Han 2007, Lothian 1999, Van Den Berg et al. 1998).

The following approach may be productive when addressing ecological, economic and social aspects of biodiversity evaluation: identification of all relevant stakeholders in the region, design of the questionnaire, where potential social, economic and environmental criteria might be identified, and presentation of this questionnaire to all stakeholders with a request to assess on a scale (e.g. from 1 to 10) the relative importance of various criteria for this particular region. Stakeholder responses could be used as a starting point to identify priorities. Then, using diverse GIS datasets, depicting various types of protected territories, species richness, information on centres of population density, number of tourists, etc., further analysis could be performed with the aim of integrating social preferences with ecological data.

Trade-Offs Between Economic and Ecological Outcomes in Biodiversity Offset Decisions

In terms of biodiversity evaluation, the past 20 years have been very productive. Anselin et al. (1989) developed one of the first overviews of how multi-criteria methods may be applied to biodiversity assessment. Margules and Usher (1981) summarised the criteria most often used in evaluation: diversity, rarity, naturalness, area, threat of human interference, representativeness, research and educational value, recorded history and potential value, etc. Goldsmith (1983) proposed a distinction between 'ecological criteria' such as size, diversity or richness and rarity, which can be more or less measured objectively, and 'conservation criteria', such as potential value and intrinsic appeal (Van Den Berg et al. 1998), which are more in the category of value judgements. Margules and Usher (1984) suggested a further separation of criteria, concluding that, for small sites, ecological fragility, threat and both species and habitat were the most important criteria, while representativeness, size, naturalness and position in an ecological/geographical unit were most important for large sites.

Multi-Criteria Decision Aid for Ecological Compensation

The field of multi-criteria decision aiding (MCDA) has developed since the 1960s. Methodological work focused on discreet methods has been carried out by Roy (Roy 1985, 1991, Roy and Vincke 1981) who pioneered the use of multi-criteria assessment with the ELECTRE family of methods. Brans (Brans et al. 1986) created the PROMETHEE method. Hinloopen and Nijkamp (1990) developed a REGIME method, while Janssen developed the DEFINITE package (Janssen 1993). Munda (1995, 2008) developed the NAIADE method. A survey of multi-criteria analysis methods is presented in (Figueira et al. 2005).

MCDA has been applied to a range of regional issues, e.g. industrial development, waste management, renewable energy. MCDA methods have also been used to analyse the issues of sustainability assessment on the macro scale.

An extensive survey of MCDA methods was offered by Guitouni and Martel (1998) and a review of several MCDA sustainability applications was undertaken by (De Montis et al. 2004). The paper by Moffet and Sarkar (Alexander Moffett 2006) presents a good overview of existing approaches to multi-criteria evaluation of biodiversity in conservation planning.

MCDA presents a new paradigm, which differs from the classical goal of finding an optimal solution subject to a set of constraints so characteristic of operations research. Within the MCDA paradigm, the primary purpose of analysis becomes a search for a compromise solution which satisfies the decision maker, rather than some illusory optimum (Guitouni and Martel 1998).

The MCDA methodological procedure can be described as a non-linear recursive process involving four steps (Guitouni and Martel 1998): (i) structuring the decision problem, (ii) articulating and modelling preferences, (iii) aggregating the alternative evaluations (preferences), and (iv) making recommendations.

Roy (2005) identifies the following basic steps in the MCDA procedure: (i) identification of alternatives; (ii) selection of the family of criteria; and (iii) choice of the 'problematic', which may be reformulated as clarification of the type of problem, the form of results, and selection of the most appropriate procedure to guide the investigation. The following types of problematic are distinguished (Roy and Bouyssou 1993):

- The choice problematic (P.α): the decision aid is oriented towards the selection of a small number of "good" actions in such a way that a single alternative may finally be chosen;
- The sorting problematic (P.β): the aid is oriented towards the assignment of each action to one category (judged the most appropriate) among a family of predefined categories;
- The ranking problematic (P.γ): the aid is oriented towards a complete or partial pre-order on A, which can be regarded as an appropriate instrument for comparing actions;
- The description problematic (P. δ): the aid is oriented towards description in appropriate language of the actions and their consequences.

A discrete multi-criteria problem can be described in general terms using the following terminology (Munda 1995):

A is a finite set of n feasible actions (or alternatives);
m is the number of different points of view or evaluation criteria g_i ($i=1, 2, \ldots, m$) considered relevant in a decision problem,
where g_i: $A \rightarrow R$ ($i=1, 2, \ldots, m$) is a real valued function representing the i-th criterion according to a non-decreasing preference,
while the action a is evaluated to be better than action b according to the i-th point of view if-and-only-if $g_i(a) > g_i(b)$.

Therefore a decision problem may be represented in a tabular or matrix form. Given the sets A (of alternatives) and G (of evaluation criteria) and assuming the existence of n alternatives and m criteria, it is possible to build an $n*m$ matrix P, called an evaluation or impact matrix, whose typical element p_{ij} ($i=1, 2, \ldots, m$; $j=1, 2, \ldots, n$) represents the evaluation of the j-th alternative by means of the i-th criterion. The impact matrix may include quantitative, qualitative or both types of information.

According to Roy (2005), the most frequently used decision aiding methods are based on mathematically explicit multi-criteria aggregation procedures (MCAP). By definition, an MCAP is a procedure which, for any given pair of potential actions, gives a clear answer to the aggregation problem. This implies:

1. various inter-criteria parameters, such as weights, scaling constants, veto, aspiration levels, rejection levels, etc., which allow us to define the specific role that each criterion can play with respect to others;
2. a logic of aggregation, which usually takes into account:
    - the possible types of dependence which we might want to bring into play concerning criteria;

- the conditions under which we accept or refuse compensation between "good" and "bad" performances[3].

Roy emphasises the significance of the logic of aggregation of the MCAP. He differentiates three types of MCAP approaches:

1. Incomparability is not allowed and the rule (aggregation function) is explicitly stated. An aggregation function might be a weighted sum, additive, multiplicative or lexicographic;
2. Incomparability is accepted, and instead of an aggregation rule, a set of tests, which focus on the conditions that must be verified for outranking, is specified. In ELECTRE methods such a set uses the concepts of concordance and discordance;
3. Primary importance is given to local judgements without considering any explicit rules of aggregation. This approach uses a formal protocol organising the interaction between the decision maker and the analyst in a logical way.

When assessing the relative importance of particular sites for the purposes of biodiversity compensation (or mitigation banking), the problematic β is the most relevant. In this case each site might be assigned to a predefined quality class, e.g. from extremely valuable, to not at all valuable, with 5–7 classes[4] in between. Therefore, a decision might be made as to which quality class a particular site belongs to and which other sites belonging to a similar class might be used as compensation (i.e. as an offset), should it be necessary to use the first site for development purposes. The MCDA method ELECTRE TRI designed to address the problematic β[5], focused on the assignment of objects to one of several predefined classes, and developed at University Paris Dauphine, might be a good candidate for such an application. The method requires explicitly defined boundaries in each criterion for each class under consideration. Other alternative methods might be considered, but a decision should be made as to which level of compensation among criteria is appropriate for biodiversity evaluation schemes, with more compensation implying weaker sustainability and less compensation implying stronger sustainability solutions. The general distinction between weak and strong sustainability is understood in the following way: more compensation among sustainability dimensions or more substitution of factors is acceptable in the case of weak, and less compensation among various sustainability dimensions or less substitution of

[3]The term "performance" is used to refer to the value of $g_i(a)$, emphasising the fact that some of the $g_j(a)$ may not have cardinal interpretations and might be defined on a purely ordinal scale. When it is useful to emphasise the quantitative nature of $g_i(a)$, the term "performance" is replaced by "valuation" (when a criterion is a gradation) or "utility" (when the criterion is a measure).

[4]The standard Likert scale is much used in various fields of research and usually comprises from four to nine points. The use of a seven to nine point scale will allow necessary quality differentiation at the same time keeping the number of categories of value manageable.

[5]Alternative methods, such as IRIS, PREFDIS, ORCLASS, and TOMASO, addressing problematique β could also be considered.

factors is possible in the case of strong sustainability. It should be noted that each of the MCDA methods requires careful tuning with the aid of a range of parameters, such as threshold levels, priorities, etc. The robustness of recommendations in this context is usually assessed using the sensitivity analysis.

Stakeholder Interviews

In order to reveal the web of stakeholder interests regarding the Nature reserve of Crau, and to create a basis for discussion of the decision criteria, several Delphi[6] type interviews with key stakeholders, involved in the consultation process regarding the management of the Crau, and supervised by the Ministry of the Environment, were arranged. Stakeholders involved in this process represent a diversity of organisations and have different goals and priorities regarding the conservation of biodiversity in the region but share an interest in the Nature reserve of Crau. The following stakeholders were approached for this survey: the Government of Provence-Alpes-Côte d'Azur (Deputy Chef de Mission), the Muséum National d'Histore Naturelle (Scientific Researcher), CDC Biodiversité, Caisse des Dépôts (Chef de projet Sud-Est), the Réserve Naturelle Coussouls de Crau (Manager of the Reserve Naturelle), Laissez-faire Farmers Association (Director). The composition of the stakeholder group is justified by the fact that they represent the key interest groups having a stake in the future of the Crau region. It was only recently that these stakeholders were gathered at the same table under the auspices of the Ministry of the Environment, and were able to negotiate important issues related to the collaborative management of the Crau region. The local residents in the area are mostly farmers and they are represented by the Laissez-faire Farmers Association. It would of course be beneficial to conduct additional interviews with the farmers directly, but budget and time constraints did not allow us to do so.

Each stakeholder was asked the same basic questions plus some additional questions unique to each stakeholder. The basic set of questions was the following:

1. What does the Crau Nature Reserve mean to you?
2. Which criteria do you think are the most important for the evaluation of different small pieces of land? (Social, Economic, Environmental)?
3. How, do you think your interests regarding the nature reserve differ from the interests of other stakeholders?

Based on the stakeholder responses, a structured list of 35 economic, social and environmental criteria was compiled (Table 9.2). In Table 9.2 each of the criteria

---

[6]Delphi method, Delphi technique – a method of using questionnaires to arrive at consensual judgements (Shorter Oxford English Dictionary). Delphi method was developed in the USA during the 1950–1960s by Project RAND (Olaf Helmer, Norman Dalkey, and Nicholas Rescher).

Table 9.2 Potential evaluation criteria, revealed by the stakeholder interviews

	Criteria/Organisation	Reserve naturelle	Museum national d'Histoire Naturelle	DIREN-PACA	Laissez-faire association	CDC Bio-diversite (Caisse des Dépôts)
	Ecological criteria					
1	Ecological habitat	V		V		
2	Presence of species	V		V		
3	Connectivity of the ecosystem	V				
4	Grass cover		V			
5	Primary production		V			
6	Soil structure and the soil biosphere		V			
	Biophysical Indicators					
7	Slope, hydrostatic behaviour of the river		V			
	Biological					
8	Specialisation of communities		V			
9	Complexity of the trophic web		V			
10	Special community index		V			
11	Bird Index		V			
12	Terrestrial trophic Index		V			
13	Leaf index		V			
14	Soil free of diseases				V	
	Economic criteria					
15	Production of lamb meat	V				
16	Benefit of agriculture		V			
17	Benefit of tourism		V	V		
18	Financial value of the land				V	V
19	Interest from the business (e.g. solar panels).				V	
20	Value of the Hay of Crau					V

(continued)

Table 9.2 (continued)

	Criteria/Organisation	Reserve naturelle	Museum national d'Histoire Naturelle	DIREN-PACA	Laissez-faire association	CDC Bio-diversite (Caisse des Dépôts)
21	Sure valuation of the land due to infrastructure activities					V
22	Costs of rehabilitation of coussoul					V
23	Value and quality of groundwater			V		
	Social criteria					
24	Social value placed on the landscape by agricultural community		V			
25	Social value placed on the landscape by the non-agricultural community		V			
26	Conflict between tree farmers and sheep farmers				V	
27	Interest in the space				V	
28	Social value of the proposed infrastructure					V
29	Patrimony value					V
30	Access to the reserve			V		
31	Participatory aspect of work and decision making			V		
32	Compliance with the government objectives of protection of biodiversity			V		
33	Preservation of the pastoral activities			V		
34	Urgency to act			V		
35	Quality of the management (management plan)			V		

was marked according to whether it was mentioned by the stakeholder or fell within the spectrum of his interests. Development of such a set of criteria is a useful first step towards a full scale MCDA of the sites in the region. It should be noted that an individual scale should be developed for each of the criteria identified. This might be a quantitative or qualitative scale, with a particular method of assessment or measurement.

Analysis

Figure 9.3 depicts in graphic form the areas of intersection between the interests of various stakeholders. It is interesting to note that, although 35 different criteria for assessment were expressed by the stakeholders, no single criterion was suggested by all stakeholders. The management of the Reserve Naturelle and DIREN PACA share an interest in preservation of the ecological habitat (1) and protection of species (2) the Museum National d'Histoire Naturelle expresses an interest in the benefit of tourism (17). Laissez-faire and the Caisse des Dépôts share an interest in the financial value of land (18). The value of undertaking a full-scale multi-criteria evaluation of biodiversity would be to account for the whole spectrum of stakeholder interests.

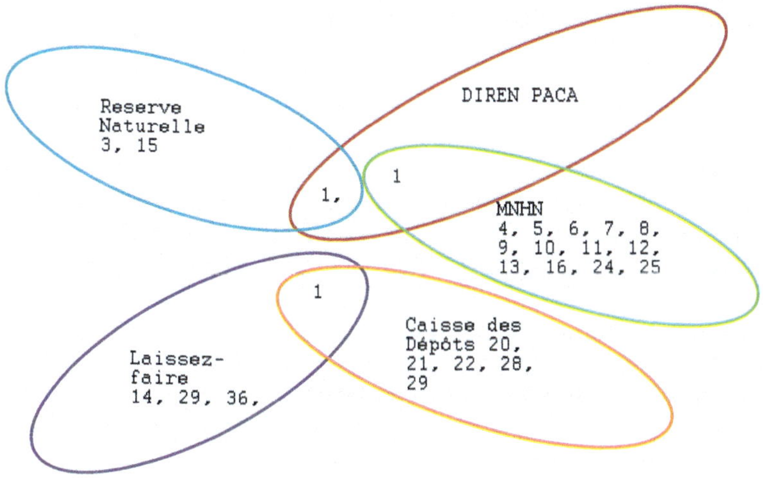

Fig. 9.3 Intersection of the sets of stakeholder interests

Discussion and Suggestions for Further Research

The analysis undertaken within this project has shown that it is possible to develop a holistic methodology which might integrate economic, social and environmental information within a multi-criteria decision aid framework to reflect the different values of particular plots of land for the purposes of ecological compensation or mitigation banking. The crucial elements of this approach are the following:

- identification of a minimal coherent set of criteria to be taken into account (extensive stakeholder consultations are required to reach consensus on which criteria should be included and the total number of criteria);
- identification of alternatives to be compared – GIS maps of the various plots of land to be evaluated need to be developed (e.g. using satellite imagery);
- selection of a multi-criteria aggregation procedure – we suggest the ELECTRE TRI method or its analogues, which are capable of assigning a range of objects (e.g. plots of land) to predefined quality classes.

Once a decision on the criteria, alternatives and aggregation procedure has been made, a multi-criteria evaluation can be undertaken, with due attention to the sensitivity of the parameters used in the procedure (e.g. threshold levels and other parameters). Full scale application of ELECTRE TRI and similar methods to the case of the Natural Reserve of Crau remains a task for future research.

Such an evaluation approach might be part of a wider system of adaptive governance, which is being created around the Nature Reserve. Following (Ostrom 2008), such a system should comply with the following five principles, which have been identified on the basis of interdisciplinary studies of failed and successful common pool resource governance systems:

1. *achieving accurate and relevant information*: the use of GIS and modern technologies, as well as building collaboration between local users, public officials, and scientific experts are identified as key elements here (the first steps in this direction have already been made in the Crau);
2. *dealing with conflict*: Ostrom highlights the idea that the possibility of conflict, which in the case of the Crau is present due to very different sets of interests among stakeholders (Fig. 9.3), should not be underestimated;
3. *enhancing rule compliance*: formal rules may become effective when participants see them as legitimate, fair, enforced, and likely to achieve intended purposes. This principle illustrates the need for extensive consultations with the stakeholders, which may ensure that the evaluation method is effective.
4. *providing infrastructure*: particular attention should be paid to the existing farmers property rights over parts of the Crau;
5. *encourage adaptation and change* (the stakeholders should be open to negotiations, be ready to adapt and legitimise change, which emerges out of friendly collaboration).

One would hope that using the principles outlined above, it might be possible to develop an effective governance system which will be capable of dealing with the contradictions highlighted in this paper.

References

Adams WM et al (2003) Managing Tragedies: Understanding Conflict over Common Pool Resources *Science* 302.5652 1915-1916

Alexander Moffett SS (2006) Incorporating Multiple Criteria into the Design of Conservation Area Networks: A Minireview with Recommendations *Diversity & Distributions* 12.2 125-137

Anselin A, Meire PM and Anselin L (1989) Multicriteria Techniques in Ecological Evaluation: An Example Using the Analytical Hierarchy Process *Biological Conservation* 49.3 215-229

Antrop M and Van Eetvelde V (2000) Holistic Aspects of Suburban Landscapes: Visual Image Interpretation and Landscape Metrics *Landscape and Urban Planning* 50.1–3 43-58

Arriaza M et al (2004) Assessing the Visual Quality of Rural Landscapes *Landscape and Urban Planning* 69.1 115-125

Brans JP, Vincke Ph and Mareschal B (1986) How to Select and How to Rank Projects: The Method *European Journal of Operational Research* 24.2 228-238

Buisson E and Dutoit T (2006) Creation of the Natural Reserve of La Crau: Implications for the Creation and Management of Protected Areas *Journal of Environmental Management* 80.4 318-326

Coeterier JF (2002) Lay People's Evaluation of Historic Sites *Landscape Urban Planning* 59.2 111-123

Costanza R (2008) Ecosystem Services: Multiple Classification Systems are Needed *Biological Conservation* 141.2 350-352

Costanza R et al (1997) The Value of the World's Ecosystem Services and Natural Capital *Nature* 387.6630 253-260

De Groot WT and Van Den Born RJG (2003) Visions of Nature and Landscape type Preferences: An Exploration in The Netherlands *Landscape Urban Planning* 63.3 127-138

de la Fuente de Val G, Atauri JA and de Lucio JV (2006) Relationship Between Landscape Visual Attributes and Spatial Pattern Indices: A Test Study in Mediterranean-Climate Landscapes *Landscape Urban Planning* 77.4 393-407

De Montis A et al (2004) Assessing the Quality of Different MCDA Methods *in Alternatives for Environmental Valuation* London, Routledge 99-133

Dramstad WE et al (2006) Relationships Between Visual Landscape Preferences and Map-Based Indicators of Landscape Structure *Landscape and Urban Planning* 78.4 465-474

Figueira J, Greco S and Ehrgott M (2005) *Multiple-Criteria Decision Analysis. State of the Art Surveys* New York, Springer

Foster J (1997) *Valuing Nature? Ethics, Economics and the Environment* London, Routledge

Gehlbach FR (1975) Investigation, Evaluation, and Priority Ranking of Natural Areas *Biological Conservation* 8.2 79-88

Goldsmith FB (1983) Evaluating Nature *in* Warren A and Goldsmith FB *Conservation in Perspective* New York, Wiley

Gorshkov VG, Gorshkov VV and Makarieva AM (2000) *Biotic Eegulation of the Environment: Key Issue of Global Change* London/New York, Springer

Guitouni A and Martel J-M (1998) Tentative Guidelines to Help Choosing an Appropriate MCDA Method *European Journal of Operational Research* 109.2 501-521

Hagerhall CM (2001) Consensus in Landscape Preference Judgement *Journal of Environmental Psychology* 21.1 83-92

Han K-T (2007) Responses to Six Major Terrestrial Biomes in Terms of Scenic Beauty, Preference, and Restorativeness *Environment and Behavior* 39.4 529-556

Harding S (2004) Food Web Complexity Enhances Ecological and Climatic Stability in a Gaian Ecosystem Model *in* Crist E and Rinker B *Scientists Debate Gaia. The Next Century* Cambridge, MA, MIT Press 255-266

Hinloopen E and Nijkamp P (1990) Qualitative Multiple Criteria Choice Analysis *Quality and Quantity* 24.1 37-56

Inhaber H (1977) Is a Land Quality Index Possible? *Water, Air, & Soil Pollution* 7.1 3-26

Janssen R (1993) *Multiobjective Decision Support for Environmental Management* Dordrecht, Kluwer Academic Publishers

Kapp, KW (1970) Environmental Disruption: General Issues and Methodological Problems *Social Science Information* 9.4 15-32

Lant CL, Ruhl JB and Kraft SE (2008) The Tragedy of Ecosystem Services *BioScience* 58.10 969-974

Lausch A and Herzog F (2002) Applicability of Landscape Metrics for the Monitoring of Landscape Change: Issues of Scale, Resolution and Interpretability *Ecological Indicators* 2.1–2 3-15

Li H and Wu J (2004) Use and Misuse of Landscape Indices *Landscape Ecology* 19.4 389-399

Lothian A (1999) Landscape and the Philosophy of Aesthetics: Is Landscape Quality Inherent in the Landscape or in the Eye of the Beholder? *Landscape and Urban Planning* 44.4 177-198

Lovelock JE (1992) A Numerical Model for Biodiversity *Philosophical Transactions: Biological Sciences* 338.1286 383-391

Margules C and Usher MB (1981) Criteria Used in Assessing Wildlife Conservation Potential: A review *Biological Conservation* 21.2 79-109

Margules C and Usher M (1984) Conservation Evaluation in Practice: I. Sites of Different Habitats in North-East Yorkshire, Great Britain *Journal of Environmental Management* 18 153-168

Martinez-Alier J, Munda, G and O'Neil J (1998) Weak Comparability of Values as a Foundation for Ecological Economics *Ecological Economics* 26.3 277-286

Millenium Ecosystem Assessment, 2004. *Ecosystems and Human Well-being: Current State and Trends, Vol. 1* Washington, DC, Island Press

Munda G (1995) *Multicriteria Evaluation in a Fuzzy Environment* Heidelberg, Physica-Verlag

Munda G (2008) *Social Multi-Criteria Evaluation for a Sustainable Economy* Berlin, Springer

Nijnik M et al (2008) Public Evaluation of Landscape Content and Change: Several Examples from Europe *Land Use Policy* 26.1 77-86

O'Neil J (1997) Managing Without Prices: The Monetary Valuation of Biodiversity *Ambio* 26.8 546-550

Ode A et al (2009) Indicators of Perceived Naturalness as Drivers of Landscape Preference *Journal of Environmental Management* 90.1 375-383

Ostrom E (2008) The Challenge of Common Pool Resources *Environment* 50.4(July/August) 8-21

Ostrom E et al (1999) Revisiting the Commons: Local Lessons, Global Challenges *Science* 284.5412 278-282

Pearce DW and Moran D (1994) *The Economic Value of Biodiversity* London, Earthscan

Peterson GL (1974) Evaluating the Quality of the Wilderness Environment: Congruence Between Perception and Aspiration *Environment and Behavior* 6.2 169-193

Regan H et al (2007) Comprehensive Criteria for Biodiversity Evaluation in Conservation Planning *Biodiversity and Conservation* 16.9 2715-2728

Roy B (1985) *Methodologie multicritere d'aide a la decision* Paris, Economica

Roy B (1991) The Outranking Approach and the Foundations of Electre Methods *Theory and Decision* 31.1 49-73

Roy B (2005) Paradigms and Challenges *in Multiple-Criteria Decision Analysis. State of the Art Surveys* New York, Springer 3-24

Roy B and Bouyssou D (1993) *Aide Multicritère à la Décision: Méthodes et Cas* Paris: Economica

Roy B and Vincke P (1981) Multicriteria Analysis: Survey and New Directions *European Journal of Operational Research* 8.3 207-218

Shmelev SE and Shmeleva IA (2009) Sustainable Cities: Problems of Integrated Interdisciplinary Research *International Journal of Sustainable Development* 12.1/2009 4-23

Taylor PD (2002) Fragmentation and Cultural Landscapes: Tightening the Relationship Between Human Beings and the Environment *Landscape and Urban Planning* 58.2–4 93-99

Tubbs CR and Blackwood JW (1971) Ecological Evaluation of Land for Planning Purposes *Biological Conservation* 3.3 169-172

Tveit M, Ode AÌŠ and Fry G (2006) Key Concepts in a Framework for Analysing Visual Landscape Character *Landscape Research* 31.3 229-255

Ulrich RS (1979) Visual Landscapes and Psychological well-Being *Landscape Research* 4 14-23

Ulrich Roger S (1986) Human Responses to Vegetation and Landscapes *Landscape and Urban Planning* 13 29-44

United Nations (2007) *Indicators of sustainable development, guidelines and methodologies* (3 rd edition) New York, United Nations

Van Den Berg AE, Vlek CAJ and Coeterier JF (1998) Group Differences in the Aesthetic Evalution of Nature Development Plans: A Multilevel Approach *Journal of Environmental Psychology* 18.2 141-157

Nunes PALD and van den Bergh JCJM (2001) Economic Valuation of Biodiversity: Sense or Nonsense? *Ecological Economics* 39.2 203-222

Van der Ploeg SWF and Vlijm L (1978) Ecological Evaluation, Nature Conservation and Land Use Planning with Particular Reference to Methods Used in the Netherlands *Biological Conservation* 14.3 197-221

Wright DF (1977) A Site Evaluation Scheme for Use in the Assessment of Potential Nature Reserves *Biological Conservation* 11.4 293-305

Zube EH, Sell JL and Taylor JG (1982) Landscape Perception: Research, Application and Theory *Landscape and Planning* 9.1 1-33

Chapter 10
Sustainable Cities: Interdisciplinary Perspective

Abstract This chapter explores the problem of integrated interdisciplinary research in the field of sustainable cities. The problematique of urban sustainability is studied in its historical and international context. Current research in the field is reviewed, and major gaps in interdisciplinary analysis are identified. This chapter makes first steps towards the development of an innovative methodology for the analysis of interdisciplinary links between various sustainability dimensions in the urban context. The dimensions taken into consideration are: sustainable energy, sustainable transport, material flows and waste management, quality of life, health, psychology of interaction with the environment, green space, biodiversity, preservation of the natural and cultural heritage, landscape architecture, ecodesign and democratic participation. Two great European cities, London and St Petersburg, are analysed from the point of view of integrative policies and instruments applied in the field of sustainable urban development. The two cities are compared using a panel of sustainability indicators.

Keywords Sustainable development • Urban sustainability • Interdisciplinary research • London • St Petersburg

Sustainable Urban Development: International Context

Urbanisation is one of the key tendencies of the twenty and twenty-first centuries. According to United Nations forecasts, between 2005 and 2030 the share of world population, living in cities, will grow from 48.7% to 59.9%. (United Nations 2005).

The growing concern of governments of developed and developing countries regarding socio-economic consequences and environmental impacts of rapid global urbanisation process resulted in a substantial programme of the United Nations activities: the first United Nations Conference on Human Settlements Habitat–I (United Nations 1976), held in Vancouver (Canada) in 1976; the resolution of the General Assembly of the United Nations on the creation of the Commission on

S.E. Shmelev, *Ecological Economics: Sustainability in Practice*,
DOI 10.1007/978-94-007-1972-9_10,

Urban Dwellings in 1977; and the foundation of the UN HABITAT centre in 1978. The HABITAT centre started to coordinate all types of activities in the field of sustainable urban development. The following activities were chosen as key: (a) policy and strategic issues in the field of the development of urban dwellings; (b) development planning for urban dwellings; (c) housing and utility services; (d) development of local communities; (e) development of infrastructure of the urban dwellings; (f) land use; (g) mobilisation of financial resources for the development of urban dwellings; (h) organisation and management of construction of the urban dwellings (UN-HABITAT 2008).

The second United Nations Conference on Human Settlements (Habitat II) held in Istanbul in 1996 (United Nations 1996b) adopted the following major documents of the UN Habitat Centre – «Habitat Agenda» – the world agenda on the sustainable development of urban dwellings, and «Istanbul Declaration», which defined the responsibilities of heads of states and governments of the countries of the world on securing the livelihood, productivity and sustainable development of urban dwellings for the whole planet.

The concept of "Sustainable development" was introduced in the report by the United Nations Commission on Environment and Development, headed by the Prime Minister of Norway, G. H. Brundtland, which bore the title «Our common future» (WCED 1987). Sustainable development is understood by the scientific community as a process of harmonious economic development, satisfying principles of social justice and environmental responsibility. Sustainable development is a compromise path between economic, environmental and social goals, between the present and the future. It should be pointed out that sustainable development is viewed as a dynamically developing concept rather than a finally formulated idea. Many researchers highlight the difficulties of precisely defining "sustainable development" as well as differences in interpretation of the notion of sustainable development in the cross-cultural context, (Fodor and Houdebine 2006, Moiseev 2000, Schmuck and Schulz 2002). The ecological economist H. Daly defines sustainable development as "a gradual social improvement without economic growth going beyond the carrying capacity of the ecosystem. Growth means increase, whereas development means improvement" (Daly 1999).

The literature review demonstrates that the sustainable cities concept emerged in the process of the general sustainability discussion. Traditionally and for a considerable period of time cities were left out of the context of debates about the means of achieving sustainability (Bulkeley and Betsill 2003). This situation can be explained by the prominence of the nature preservation concept, which dominated in the field of environmental thought from the beginning of the nineteenth to the end of the twentieth century. In this context, rural and wild nature were associated with the environment, which needed to be protected from advancing cities and culture. Therefore, throughout the last century the environment was seen as something external to the city. Efforts were made to integrate environmental issues within the scope of general urban design and management problems only with the mass introduction of water and sewage systems, creation of parks and recreation zones, and the development of land-use planning traditions.

In the Brundtland report mentioned above, these issues received support in the form of a special chapter, devoted to urban environmental problems. It was underlined that, because the majority of the world population will live in cities in the future, cities should be the central focus of deliberations concerning sustainable development.

The Expert Group on the Urban Environment had a special task to support the cities in advancing the sustainable development concept. However, only after the conference in Rio de Janeiro in 1992 were cities recognised as an important sphere of application of sustainable development ideas. Then the initiative was passed to local municipal authorities under the auspices of the International Union of Local Authorities (IULA)) and World Federation of United Cities (WFUC), which promoted local community initiatives. The Chap. 28 of the Agenda 21, a global plan of action to reduce human impact on the environment (United Nations 1992) contained an appeal to national and local governments to assist in the implementation of sustainable development principles on a local scale. At local community level this provided an impetus for the development of independent initiatives, as in London (London 21 1998). Without a doubt, the Conference in Rio de Janeiro had two important consequences for defining the role of cities: the fact that sustainable urban development is, to a great extent, a solution to environmental problems, and that there is a great potential for international cooperation among between as e.g. in the Baltic Sea region. Agenda 21 was supported by the Habitat II Programme.

The special 25th session of the United Nations General Assembly "Istanbul+5", devoted to the implementation of the "Habitat Agenda" took place in July 2001 in New York. This session confirmed that during its 5 years of operation, through the development of international cooperation, UN Habitat had achieved significant results in improving the conditions of life of poor social groups. The Assembly unanimously adopted the «Declaration on Cities and Other Human Settlements in the New Millennium» (United Nations 1996b), in which the principles and goals of sustainable urban development were outlined. The status of the Habitat Centre was raised to the level of the United Nations Habitat Programme in 2001, headquartered in Nairobi, Kenya.

Strategic plans for sustainable development and Agenda XXI were discussed at the World Summit for Sustainable Development in Johannesburg in 2002 (United Nations 2002) and at subsequent international meetings. For example, during the conference "The City of the XXI Century" (REC 2000) the strategy for improvement of quality of life in the city, including environmental, cultural, political, institutional, social and economic components was included as a key element of the urban sustainable development strategies.

The European Union considers sustainable urban development to be one of the priority directions of its activities, a Thematic strategy on Urban Environment was adopted by the commission in 2006 (European Commission 2006).

Sustainable City: Formulating the Problem

The development of a methodology for the study of the city and management of the city as a complex multi-layered holistic system poses certain difficulties. We share the point of view of many authors, conducting research in this field: (Girardet 2004, Hall 2006). The methodological starting point of such approaches is the move away from a mono-disciplinary approach, where the complex multi-dimensional system is split into a multitude of separate objects, each of which requires an individual method of enquiry, analysis and management. The systems analysis theory suggests that the behaviour of complex systems is often characterised by emergent properties, appearing as a result of the interaction of the elements or subsystems of the system under consideration. Therefore an interdisciplinary approach is capable of providing a qualitatively novel understanding of the problems of urban development and management, determined by multidimensional synthesis. Methodologically, the systems approach seems to be the most productive, allowing us to conceptualise the multidimensional interdependencies among the most relevant components of the object under consideration. The ideal and real model can be substantially different in terms of the quality, complexity and the direction of links.

The literature on the integrative sustainable urban development problems is substantial. A review of the literature on sustainable cities reveals the following main problem areas (Fig. 10.1): *energy* (Boyle 2004, Nijkamp et al. 1999); *sustainable transport* (Banister 2005); *material flows and urban metabolism* (Swyngedouw and Heynen 2005); *landscape architecture and design* (Carmona et al. 2003, Nefedov 2002); *green space* (Clark 2006, Rees 2002); *economic activity and planning* (Banister et al. 1999); *environmental design and architecture* (Orr 2004); *psychology of interaction with the urban environment* (Chernoushek 1989, Gifford 2002, Lynch 1960, Milgram 1974, Shteinbach 2004); *democratic participation in the decision making process* (ECOM 2005), *cities in the context of globalisation* (Eade and Mele 2002, Hall and Pfeifer 2004), as well as the studies using *integrative approach* focusing on two or several urban sustainability problems: (Hall 2006, Kinzing 2005, Ravetz 2000).

Figure 10.1, based on frequencies of keyword occurrence in the Scopus citation database, presents the distribution of studies on sustainable cities according to the dimensions highlighted above. It is interesting to note that although analysis of planning and water management issues together with energy, transport, waste management and health have been the dominant focuses of attention, there has been relatively less attention devoted to governance, material flows, quality of life, environmental psychology, green space, and natural and cultural heritage.

A brief review of a number of significant publications on sustainable cities will be presented further. (Balocco et al. 2004) in their research use the concept of entropy and energy indicators for the study of sustainability of cities; Bennet and Newboroough (2001) suggest the methodology of an energy audit of cities and regions within the framework of national and international obligations to reduce CO_2 emissions and implementation of the Agenda 21; Brownsword et al. (2005)

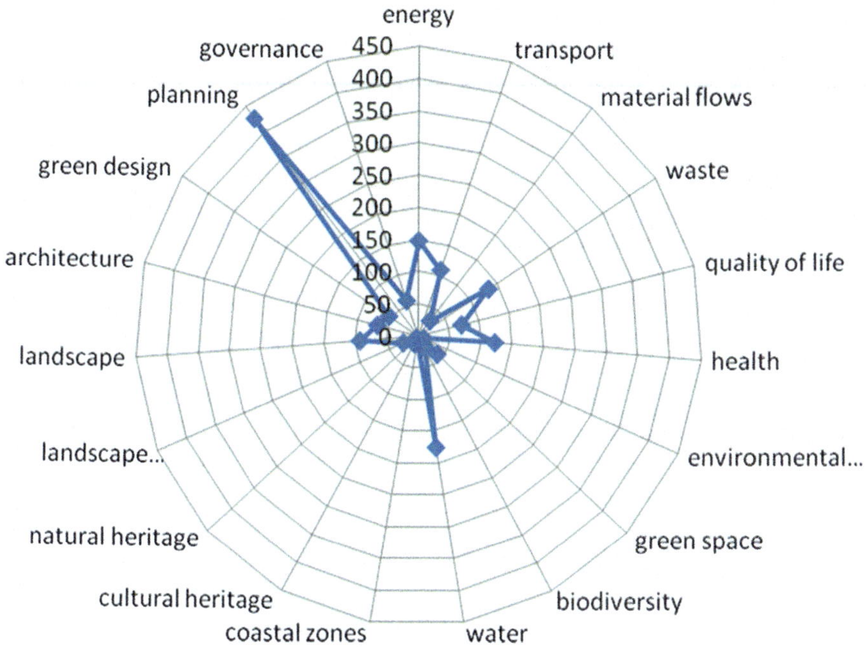

Fig. 10.1 Sustainable cities literature devoted to different sustainability dimensions (based on precise keyword search, January, 2009) (Source: www.scopus.com)

consider the results of the analysis of energy demand on the scale of the city of Leicester (UK). An optimisation model of energy use is being created taking into account measures for improving energy efficiency and use of renewable energy technologies. Nijkamp and Pepping (1998) consider the contribution of energy policy towards sustainability in urban systems. A comparative analysis of the use of renewable energy technologies in large (Amsterdam, Rotterdam, Milan, Turin, Thessaloniki, Chania) and small cities of Italy, the Netherlands, and Greece is made. Muniz and Galindo (2005) analyse the determining factors of the environmental impact of passenger movements among the different districts of the Barcelona metropolis. It is shown that urban planning plays a significant role in the difference in ecological impact. (Rees 2003) considers ecosystem principles of the analysis of cities and pays particular attention to the concept of ecological footprint (Wackernagel and Rees 1996) and its use at city level. However, our experience of its application in the German context showed its artificial character, the limitations and the need for deeper methodological justification (Giljum et al. 2007). Nijkamp et al. (1997) apply the spider-web model to the analysis of urban transport systems. Eight dimensions of sustainability of transport systems are considered, including spatial, institutional, economic and socio-psychological factors and the key management impacts in the system are analysed. Several scenarios of the development of the transport system are considered and the total CO_2 emissions from the functioning of the system are assessed. The need to strengthen the role of public transport is underlined.

(Hammer and Giljum 2006) analyse material flows on the scale of the city and study material balances of such cities as Hamburg, Vienna and Leipzig. Leach (1997) offer a model for material flows analysis in an urban system (using the example of paper flows). The authors use the method, integrating life cycle analysis and input-output analysis. (Camagni et al. 1998) consider positive and negative externalities in urban systems in static and dynamic settings as well as strategies for improving urban sustainability. (Fung and Kennedy 2005) develop an econometric model, linking parameters of urban metabolism with economic variables. This model is applied to the analysis of the Toronto metropolis (Canada). The authors reviewed different modelling methods, used in urban development analysis- regional economic models (including econometric and input-output), land use models, urban metabolism models, spatial ecological models. The results of the modelling experiments focused on CO_2 are analysed; (Holden and Norland 2005) consider the connection between land use characteristics in cities and energy use in transport and in households. Their results support the thesis on the advantages of the compact city as a sustainable urban form using Oslo (Norway) as an example.

Finally, there is a group of articles using the integrative approach. First, a paper by (Button 2002), devoted to the problem of choosing environmental, economic and social indicators for the analysis of urban development should be mentioned. It considers market, political and administrative feedback loops in the dynamic system of urban development. Suggests a set of indicators for the evaluation of urban or regional sustainability, and uses an interdisciplinary approach in his overview, which is based on the principles of integrated assessment. (Shane and Graedel 2000) developed a method of representation of valuation results based on a set of urban sustainability criteria. The proposed criteria-set is applied to the analysis of the city of Vancouver (Canada). (Wiek and Binder 2005) offer a new approach to the evaluation of the sustainability of cities and regions, based on taking the three main shortcomings of the existing studies into account – the use of unrelated indicators, lack of coherent analysis of goals, neglect of interdisciplinary approaches. Special emphasis in the article is put on the synergic and contradictory consequences of decisions taken. In our opinion, this paper gives the most comprehensive systems overview of the problem of urban sustainability.

Undoubtedly, publications covering real problems and describing the real experience of the realisation of Local Agenda 21 in different cities of the world are of great interest (Bowin 1997).

An important place among urban studies is occupied by psychological and social aspects of city development and functioning. Works by K. Lynch, which appeared in 1960s (Lynch 1960) stimulated a large number of studies on the cognitive and mental representation of cities, and research on psychological mapping of such large cities as Paris and New York (Milgram 1974), perception of architectural space, and urban design, taking environmental psychology principles into account (Churchman 2002). Research by Prochansky (Proshansky 1978, Proshansky et al. 1983) served as a stimulus for the study of urban stress, and led to the analysis of place identity and attachment to place of residence (Gouveia 2002). The works by (Barker 1968) and his idea of «behavioural setting» in psychology – the environment

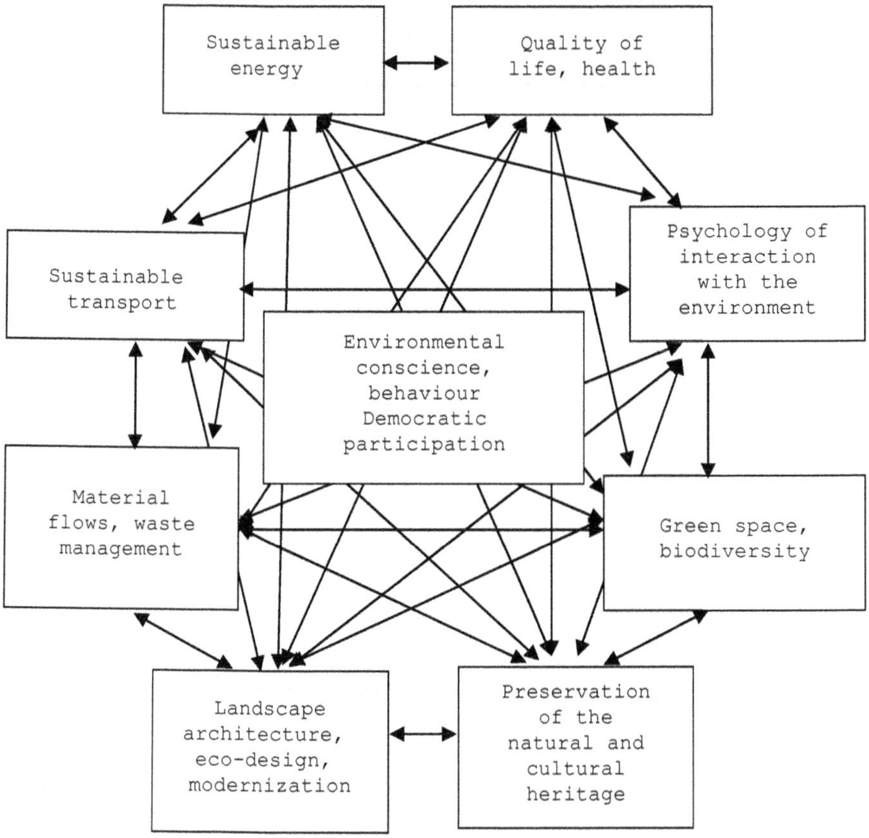

Fig. 10.2 The key problem areas in the field of sustainable urban development

in which the behavioural model is observed – served as a basis for the study of problems of behaviour in the urban environment, antisocial and aggressive behaviour, social helplessness and mutual help, the problems of neighbourhood and local communities, the problems of quality of life in cities and perception and understanding of the city as a holistic social organism (Gifford 2002). Undoubtedly, publications where the psychological aspects of the realisation of the Local Agenda 21 are analysed, are of great interest, since they explore attitudes towards the problem of sustainable development at the local level (Lindström 2003, Schmuck and Schulz 2002). Taking into account the fact that most environmental problems in cities are created and solved by the people, the study of the psychology of the environmental conscience and behaviour towards the environment (Kaufmann-Hayoz 2006, Medvedev and Aldasheva 2001, Shmeleva 2006) seems to be extremely relevant.

Generalisation of the results of the published research, fulfilled projects, as well as outcomes of conferences and seminars (Shmeleva and Shmelev 2007) allows us to distinguish the key problem areas of the interdisciplinary description of the city as a sustainable development object (Fig. 10.2).

A "sustainable city" is a concept characterising the development of the city as a holistic system, in which social, economic, environmental and institutional aspects of development are harmoniously integrated. Our vision of the main aspects of sustainable development of a city in their interrelation and interdependence is depicted in the form of a graph (Fig. 10.2). The nodes of the graph represent the major problem fields. The arrows connecting the nodes highlight the links of influence existing among the problem areas. For example, development of a sustainable public transport system in a city will improve the quality of life and public health, reduce impacts on green space and biodiversity, reduce total energy requirements and help to preserve natural and cultural heritage sites due to avoidance of emissions.

A city is a complex system, in which a multitude of objects and processes are interlinked in time and space. The houses where people live, enterprises where people work, museums, sport halls, restaurants, cafés, were people spend their free time are connected in space by the infrastructure of roads, energy and information networks. All enterprises, universities, restaurants, banks use energy and material resources; energy and material resources are also spent on transportation of goods and services and transfer of citizens from the places where they live to the places where they work and back, on heating buildings, lighting houses and courtyards, streets and avenues. A city is a live, dynamic organism, which every second consumes water and energy, and produces liquid, solid and gaseous biological and chemical wastes, creating an impact on the environment.

Different types of spatial organisation (streets, squares, avenues, regional centres, multi-storey housing, low-storey housing, parks and gardens), composition of various methods of energy generation (nuclear, wind, solar, gas, coal), different ways of organising the transport system (priority to private or public transport, trams, railways, aviation), energy saving in houses and enterprises, recycling of waste as opposed to landfilling and incineration, creation of conditions for the development of trade, leisure, sport activities, play a key role in determining the ecological impacts of the city on the environment on the one hand and the impacts of the environment on the citizens on the other.

A city is a social and information environment, where communities reside, connected in social and information networks. Citizens, independent of their age, ethnic and professional group, social status, incomes, and so on would like life in the city to be happy, comfortable, healthy, and safe, and they would like the environment surrounding them to be convenient, aesthetically pleasing and environmentally clean. Participation in decision making on matters of urban development (democratic participation) gives the citizens a feeling of attachment to their city, a place identity, a sense of belonging to the community of fellow citizens and opportunities for self-realisation. Experience shows that when a city administration actively follows the principles of sustainable development, it can offer opportunities and create conditions for satisfaction of these needs. When it does not happen, conflict situations emerge, which may often be focused on the destruction of historic city centres, destruction of parks and gardens, and excessive construction.

A city is history, the place where our ancestors lived, and our descendants will live. It is a space, where the landscapes and architectural monuments of the past are present. But, as a human organism, the city gets older; its monuments require care and effort for their preservation. The city space is a space, where old and new interact, which can become a space of contradictions or a space of harmony. The city is a repository of the cultural heritage of mankind, with its museums, archives, theatres, exhibitions, concerts, libraries, universities, and scientific institutes.

The city is a geographical environment, geology, landscape, bodies of water (rivers, lakes, seas, oceans, coastal areas), weather, climate, nature – fauna and flora, ecosystems, biodiversity, parks and gardens, forests, birds and insects, domestic, wild and homeless animals and even natural resources deposited in the ground.

Acting as a dynamic economic and political agent, the modern city is linked with the infrastructure of the whole region, country, continent and the world as a whole. Its economy is linked with the economies of other regions and countries, fuel, food and other goods are delivered to the city from the region, supporting it and other areas, goods produced in the city are shipped to other regions and countries. The sectoral structure of the city's economy determines the quality of the impacts of the city on the environment and allows us to find the sensitivity points; by influencing these we may be able to stimulate a more environmentally friendly development.

Therefore, this description allows us to represent the city as a complex multidimensional object of study and management from the sustainable development point of view. It should be said that the detailed description, quantification and further conceptualisation of links existing between the problem fields identified remain an important field for future research.

Sustainable Development Strategies for a Large City: London

The capital of Great Britain, London has unique experience of strategic planning for sustainable development and is one of the leaders among large European capitals in this field. The initiatives of the London administration are supported by Londoners and the national government. The former Mayor of London, K. Livingstone acted as an initiator of many strategic changes. The London Assembly undertakes substantial efforts to encourage democratic participation, popularise ideas of sustainable development of the city and clarify its policy to citizens. Recently, a whole range of "Environmental strategies" was adopted by the Mayor of London, which acted as a framework for formulating tasks to develop a plan of actions with the aim of making London a cleaner, greener and more sustainable city.

These strategies are as follows:

- Air Quality Strategy;
- Ambient Noise Strategy;
- Biodiversity Strategy;
- Energy Strategy;
- Municipal Waste Strategy;
- Business Waste Management Strategy;

- Economic Development Strategy;
- Culture Strategy;
- Transport Strategy;
- Food Strategy;
- Housing Strategy
- Water Strategy

The Health Inequality Strategy is being developed.

We will consider some of the strategies listed and the results of their implementation in more detail. The recently introduced congestion charge may be seen as an example of an interdisciplinary solution to the problem, as it affects four problem areas at the same time: sustainable transport, quality of life and health, material flows and psychology of interaction with the environment. This charge produced results that cannot be ignored: the number of cars in the area, where the charge has been operating, was reduced by 26%, CO_2 emissions went down 16.4% from 2002 to 2003, PM_{10} – by 15.5%, NO_x – by 13,4%. The funds obtained through exercising this tax were channelled towards the modernisation of the public transport system, which relieved psychological tension which emerged after the introduction of this tax.

According to the results of monitoring and research, the main sources of air pollution in London are cars and aviation, and to a lesser extent buildings and industry. The air quality in London (GLA 2002a) is regulated in the following way: the UK government has set targets to reduce the emissions of nine major pollutants. The concentrations of seven of them, which are nitrous oxide (NO_2), particulate matter (PM_{10}), sulphur dioxide (SO_2), carbon monoxide, benzene, 1,3-butadiene and lead have to be addressed at the local level, and the two others – ozone and polycyclic aromatic hydrocarbons at the national level and at European Union level. National reduction goals determine the average levels of concentration of such substances in the atmosphere at different time periods (for example annually or daily, depending on health impacts), which need to be achieved to a particular deadline. According to UK laws, the Mayor of London is responsible for the reduction of emissions of the seven local pollutants to achieve the targets, set by government.

The key priorities in the field of transport, according to the Transport strategy of London (GLA 2001), are: reduction in the number of traffic jams, increase in construction investment, redesign and maintenance of the underground system, radical improvement of the bus transport system, improvement of integration of the regional railway trains into the overall London transport system, improvement in the carrying capacity of the transport system – creation of public transport lanes, creation of rules for buses to stop on demand within designated stopping zones, reduction of dependency on private transport by offering the whole spectrum of opportunities for the use of public transport, development of a system of electro transport (e.g. trams as an environmentally friendly means of transport), support of local initiatives, including walking and cycling, improvement in the reliability and effectiveness of transporting goods and services, thereby reducing environmental impacts at the same time; improved accessibility to transport in London, including for the people with disabilities, as well as the increasing the integratedness of the whole transport system – development of transfer stations, improvement in information support and improvement in the state of stops.

The goals of the energy strategy of London (GLA 2004) are the reduction of impact of London on global climate change by minimising CO_2 emissions in all sectors (commercial, household, industry and transport) by improving energy efficiency and using renewable energy sources, reduction of "fuel poverty", securing access to heating even for the least privileged social groups, as well as supporting the contribution of the energy system to the economy of London by increasing employment and promoting innovations in the field of energy technologies.

The Municipal Waste Management Strategy (GLA 2003) has as its main goals changing the lifestyle of Londoners, aimed at minimisation of the generation of waste and reduction in environmental impacts. Reduction, reuse and recycling of waste are announced as the main priorities until 2020. According to statistical data in 2001/2002 about 73% of municipal solid waste was sent to landfill and only 8% was recycled. According to the strategy the large-scale separate waste collection is planned, as well as an increase in landfill tipping fees.

The London biodiversity strategy (GLA 2002b) includes 14 activity directions, among which are preservation, the management and improvement of biodiversity in London, the organisation of the network of waterways, rivers and canals, "The Blue Ribbon"; the improvement and creation of new parks, improvement of access and stimulation of interest in nature; support of farming traditions and biodiversity in agriculture in the nearest suburbs, encouragement for the greening of the city, support for ecological education, popularisation of species of plants and animals living in London, preservation of existing parks and green areas in the central part of London and so on.

Apart from planning aimed at the functioning of the system as a whole, London has many independent projects, related to sustainable development at the local level, for example the building of new energy-efficient office buildings or living quarters and construction using the principle of minimisation of current CO_2 emissions related to their functioning. The BedZed complex in West London might be an example of such development.

One of the priority tasks of the sustainable development of London for over 20 years has been the task of stimulating the activity of local communities (residents of municipalities of London) in the implementation of Local Agenda 21, involving citizens and other stakeholders in decision making at the level of local administration, the development of network interaction within London 21 Sustainability Network (Jopling 1999).

Sustainable Development of St Petersburg: Between Past and Present

Saint Petersburg is the second largest Russian city, the fourth largest city in Europe after Moscow, Paris and London. It has been called the «Northern Venice», «Magnificent St Petersburg», one of the slightly over a dozen European cities, the whole central part of which is on the UNESCO World Heritage list. Its geopolitical and cultural importance in the context of European development is very significant. (Yagya 2005).

The main goals of development of St Petersburg for 2005–2025 are defined in the General Plan (LASP 2005). The main development goals according to the General plan are the ongoing improvement of the quality of life of all strata of the population of St Petersburg with intention of securing European standards of living, the development of St Petersburg as a multifunctional city, integrated into the Russian and world economy, providing a high quality business environment, and strengthening St Petersburg as the main Russian contact centre of the Baltic Sea region and the North-West of Russia. The goals of territorial planning in St Petersburg are: securing the sustainable development of St Petersburg; improving the quality of the urban environment, preserving and regenerating the historical and cultural heritage; developing engineering, transport and social infrastructure; securing taking into account the interests of the Russian Federation, the interests of the citizens of St Petersburg and their groups, and interests of intra-city municipal units in St Petersburg.

The General Plan of development of St Petersburg implies the development of a whole range of local St Petersburg laws, aimed at regulating the main fields of the city's development:

- On the cultural heritage sites (historical and cultural monuments) in St Petersburg, including documents, regulating the preservation of the centre of St Petersburg as a UNESCO world heritage site;
- On the natural-healing resources, medical-recreational sites and resorts
- On specially protected natural territories;
- On the Earth's Interior;
- On Soils;
- On Waste Management;
- On Forests;
- On Fauna;
- On nature management and environmental protection;
- On the Preservation of Air Quality;
- On Protection from Noise;
- On Radiological Safety;
- On Electromagnetic Safety.

Despite the fact that sustainable development is denoted as a priority goal of the development of St Petersburg, it should be mentioned that in the list of the "priorities of socio-economic development" listed under the heading "The Goals of Territorial Planning" there are no environmental goals, the majority of the listed priorities relate to the development of certain sectors of industry, trade, science, and the commercial sector. The General Development Plan is a cause of big debates and much resentment according to the press, it seems that the main dimensions of sustainable development are not linked in it at all; the key concepts on which the development of St Petersburg is based, according to the city Administration are stability, balance, reconstruction and organic growth, whereas the non-financial components of quality of life, democratic governance in decision making, and reduction of environmental impacts are not listed as key priorities. Given the current priorities one can expect a further increase in pressure on the environment from industry and transport. The speed, coordination and degree of the planned innovation in the area of public transport and the organisation of ergonomic, safe and human-friendly living space seem to be insufficient.

At the same time, monitoring of the quality of the environment is constantly carried out by the Environmental Protection and Ecological Safety Committee of the Administration of St Petersburg. The annual report on the quality of the Environment in St Petersburg is published regularly (ASP 2005). Several years ago an international project on «Information and Communication Technologies to Strengthen Sustainable City Management» was started; this was focused on the creation of an interactive information system, which could help decision makers to receive information on the concentration of pollutants, total emissions, the quality of the green areas, the generation of waste and other spatially distributed data.

The situation in St Petersburg today from the citizen's point of view is characterised by the following: development of the transport system cannot keep up with the development of the city, traffic jams have become an inherent part of urban life, construction of much needed new underground lines is delayed for several decades, tramway routes are being demolished throughout the city to give priority to private transport, public transport is not seen as a priority, there is no system for regulating parking on major city streets, there are no cycle paths inside modern districts. The satisfaction of the immediate economic interests of building companies leads to the destruction of green areas: parks, trees in the streets, and green spaces are deliberately taken out of Park Management control and given to developers to create multistorey buildings, with ecological expertise often applied in a very superficial way, and the opinions of citizens largely ignored by the administration. The historic centre is being slowly destroyed by the careless construction and renovation process. All these factors adversely affect the quality of life of the citizens, pose a threat to their health, and deepen their psychological stress and discomfort. It is obvious that the solution to these problems requires their consideration as the public goods management problems and the problems of interaction of the human being and the environment. On this basis, taking a systems interdisciplinary approach, evolutionary strategies for the development of St Petersburg might be formed. In our opinion it is not too late to change the vector of the evolution of St Petersburg as a European city towards sustainability.

Sustainable Development Indicators: Two Case Studies

The set of sustainable development indicators, created for the UN Commission on Sustainable Development, was first published in 1996 (United Nations 1996a). Then, in 2001 it was reconsidered and included four main groups of indicators – economic, environmental, social and institutional. In 2005 the restructurizing of the set of sustainable development indicators took place, and 50 core indicators were chosen out of 96 present in the previous edition (United Nations 2007). As a result, a decision was made that the key indicators should satisfy the following criteria: (1) address problems which relate to the sustainable development process in the majority of countries, and not in a single country; (2) give useful information, not available from other indicators (3) provide an opportunity to calculate them for the majority of countries on the basis of the available data.

Respectively, indicators in the new edition are not subdivided into the four groups named above, which, in our opinion, reflects the interdisciplinary nature of the sustainable development idea and an understanding of the deep connections between processes of varying nature, taking place in complex open systems. This corresponds to V. Vernadsky's idea proclaimed in the middle of the twentieth century that science should specialise in problems rather than in scientific disciplines (Vernadsky 1936).

At the moment all indicators are grouped in 14 thematic categories, each of them containing subcategories, for example in the "Atmosphere" thematic category, subcategory "Climate Change", we find the indicator "CO_2 emissions" and so on. Therefore, all indicators are joined in a "goal tree", which allows to quantify them with the help of available statistical parameters. These divisions are as follows:

- Poverty
- Governance
- Health
- Education
- Demography
- Natural hazards
- Atmosphere
- Land
- Freshwater
- Oceans, Seas, and Coasts
- Biodiversity
- Global Economic Partnership
- Economic Development
- Consumption and Production Patterns

Drawing on the available regional sustainability statistical data, we will try to compare the two cities, and explore the major differences (Fig. 10.3). The sustainability indicators covered in this comparative analysis are: total regional GVA, unemployment, total consumption of electricity and gas, representing the economic dimension; total population, life expectancy at birth for males and females, and recorded crimes, representing the social dimension, and total area,% of green space, GHG emissions: CO_2, NH_4, NO_x, precursors of acidification: SO_2, NH_4, and transport related emissions CO, PM_{10}, NMVOC, representing an environmental dimension. It should be noted that some data is not collected by the statistical authorities and is not available (e.g. the weight of generated MSW in St Petersburg). Figure 10.3 highlights the similarities and differences between the two great cities.

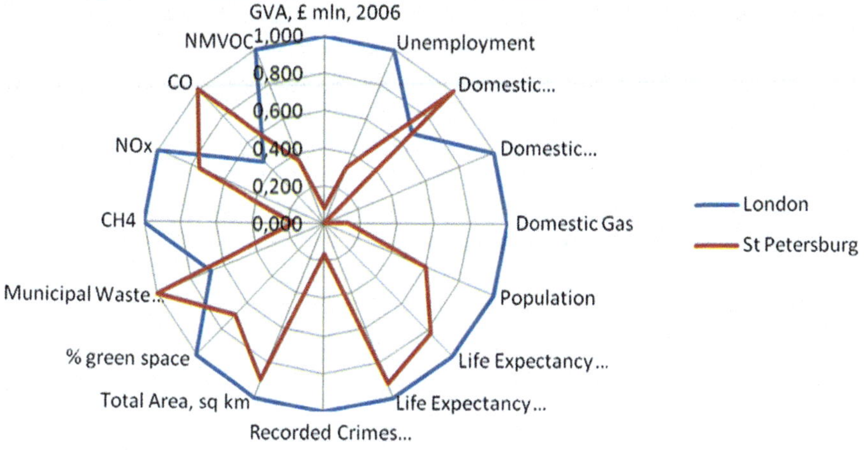

Fig. 10.3 Comparative sustainability analysis of London and St Petersburg

It is interesting to note, that although the total area of London and St Petersburg is very similar (1,596 and 1,439 sq. km respectively), the population of London being around 65% higher, the economic output of London is 12.5 times higher than that of St Petersburg, Londoners on average live longer lives (the difference in male life expectancy is more than 13 years). By contrast, St Petersburg has much less unemployment and recorded crimes. Both cities have remarkably similar rates of availability of green space per capita (80.25 and 82.95 m^2 respectively). Electricity consumption in St Petersburg is slightly higher and gas consumption is lower, since hot water and heating are supplied centrally. The emission profiles of the two cities are characteristically different. London generates considerably higher emissions of CH_4 (by a factor of 5.9), produces 32% more NO_x emissions, and NMVOC (by a factor of 2.7). However St Petersburg generates considerably more CO (by a factor of 2.1), and municipal solid waste (according to unofficial data).

Conclusion

Large cities in Europe experience similar problems due to the need to supply growing amounts of energy to satisfy increasing demands, to manage increasing flows of municipal solid waste, and to tackle emissions from the expanding car fleet. Solutions that have been found historically – development of public transport systems, waste recycling, a change in lifestyle, including teleworking, the intelligent design of energy systems and the shift towards renewables are implemented to a different extent in large European cities, with some being leaders in one field and some in another. It is interesting to see how the populations of the two cities can produce considerably different economic value, which can perhaps be explained by London being the financial capital of Europe and a centre of governmental activities.

The tendency of the St Petersburg population to live shorter lives is worrying, and one explanation might be increased work pressure, a lack of a work/life balance, low levels of pay, lower levels of ambient air quality and background stress, caused by the inefficiencies of the public transport system design. Both cities have developed a set of policy instruments to tackle wider sustainability issues, with St Petersburg relying on the City Development Plan and local laws and London basing its policies on a set of sustainability strategies. It seems that the legal responsibility of the Mayor of London for reducing concentrations of certain emissions and the lack of such obligations for the Mayor of St Petersburg play a role in the levels of success in addressing sustainability issues. Also, thematic strategies in London allow more flexibility and opportunity to adjust to changing conditions than the relatively more rigid legal framework in St Petersburg. Democratic participation in the governance process in St Petersburg is at a much lower level than in London. A holistic approach to urban sustainability incorporating the diversity of perspectives: the overview of the extent of research in the relevant problem areas to identify knowledge gaps, the analysis of interdisciplinary links among different sustainability dimensions to reveal the problem complexity and hidden connections, and comparative sustainability assessment to highlight differences in performance and policies used, create a solid methodological basis for further detailed investigation.

References

ASP (2005) Administration of St Petersburg, Ochrana Okruzhajuschej Sredy, Prirodopolzovanije I Obespechenije Ekologicheskoj Bezopasnosti v Sankt Peterburge

Balocco C et al (2004) Using Exergy to Analyze the Sustainability of an Urban Area *Ecological Economics* 48.2 231-244

Banister D (2005) *Unsustainable Transport: The Transport Crisis* New York, Spon Press Routledge

Banister D, Button K and Nijkamp P (1999) *Environment, Land Use and Urban Policy* Cheltenham, Edward Elgar

Barker RG (1968) *Ecological Psychology: Concepts and Methods for Studying the Environment of Human Behavior* Stanford, Stanford University Press

Bennett M and Newborough M (2001) Auditing Energy Use in Cities *Energy Policy* 29.2 125-134

Bowin K and Magnusson S (1997) *Local Initiatives for Sustainable Development* Stockholm, Naturskyddsforeningen

Boyle G (2004) *Renewable Energy* (2nd edition) Oxford/New York, Oxford University Press in association with the Open University

Brownsword RA et al (2005) Sustainable Cities – Modelling Urban Energy Supply and Demand *Applied Energy* 82.2 167-180

Bulkeley H and Betsill M (2003) *Cities and Climate Change: Urban Sustainability and Global Environmental Govenance* London, Routledge

Button K (2002) City Management and Urban Environmental Indicators *Ecological Economics* 40.2 217-233

Camagni R, Capello R and Nijkamp P (1998) Towards Sustainable City Policy: An Economy-Environment Technology Nexus. *Ecological Economics* 24.1 103-118

Carmona M et al (2003) *Public Places – Urban Spaces: A Guide to Urban Design* Oxford, Architectural Press

Chernoushek M (1989) *Psychologija Zhiznennoj Sredy*, Moscow, "Mysl"
Churchman A (2002) Environmental Psychology and Urban Planning: Where Can the Twain Meet? *in* Bechtel RB and Churchman A *Handbook of Environmental Psychology* New York, Wiley 191-201
Clark P (2006) *The European City and Green Space: London, Stockholm, Helsinki and St Petersburg, 1850–2000* Aldershot, Ashgate
Daly HE (1999) *Ecological Economics and the Ecology of Economics*Northampton, Edward Elgar, Cheltenham, UK and Northampton, MA, USA
Eade J and Mele C (2002) *Understanding the City: Contemporary and Future Perspectives* Oxford/ Malden, Blackwell Publishers
ECOM (2005) *Konferenzii Grazhdan. Novyje Vozmozhnosti Uchastija v Prinyatii Reshenij* St Petersburg, Centr Expertiz ECOM St Petersburgskogo Obschestva Estestvoispytatelej
European Commission (2006) Thematic strategy on the urban environment Available online at: http://ec.europa.eu/environment/urban/thematic_strategy.htm
Fodor F and Houdebine A-M (2006) The Notion of Sustainable Development in Various Languages *International Journal of Ecological Economics and Statistics* 5.2 113-118
Fung M and Kennedy CA (2005) An Integrated Macroeconomic Model for Assessing Urban Sustainability *Environment and Planning B: Planning and Design* 35 629-656
Gifford R (2002) *Environmental Psychology: Principles and Practice* Colville, Optimal Books
Giljum S et al (2007) *Scientific assessment andevaluation of the indicator 'Ecological Footprint'* Report for Federal Environment Agency (UBA, Government of Germany)
Girardet H (2004) *Cities People Planet: Liveable Cities for a Sustainable World* Chichester, Academy Press
GLA (2001) *The Mayor's transport strategy* London, Greater London Authority
GLA (2002a) *Cleaning London's air: Highlights of the Mayor's air quality strategy* London, Greater London Authority
GLA (2002b) *Connecting with London's nature. Highlights of the Mayor's biodiversity strategy* London, Greater London Authority
GLA (2003) *Rethinking rubbish in London. Highlights of the Mayor's municipal waste management strategy* London, Greater London Authority
GLA (2004) *Green light to clean power. Highlights of the Mayor's energy strategy* London, Greater London Authority
Gouveia V (2002) Self, Culture, and Sustainable Development *in* Schmuck P and Schultz WP *Psychology of Sustainable Development* Boston, Kluwer Academic 151-174
Hall P (2006) *The Polycentric Metropolis: Learning from Mega-City Regions in Europe* London, Earthscan
Hall P and Pfeifer U (2004) *Urban Future 21, A Global Agenda for Twenty-first Century Cities Federal Ministry Transport Building and Housing* London/New York, Sponspor Press
Hammer and Giljum (2006) NEDS Working Paper No. 6, Hamburg Shmeleva, I.A. Psychologija Ekologicheskogo Soznanija, St Petersburg University Press, St Petersburgg
Holden E and Norland IT (2005) Three Challenges for the Compact City as a Sustainable Urban Form: Household Consumption of Energy and Transport in Eight Residential Areas in the Greater Oslo Region *Urban Studies* 42.12 2145-2166
Jopling J (1999) *London Pathways to the Future. Thinking Differently* London, Central Books
Kaufmann-Hayoz R (2006) Human Action in Context: A Model Framework for Interdisciplinary Studies in View of Sustainable Development *Umweltpsychologie* 10.1 154-177
Kinzing A (2005) Urbanization: The Coming Global Challenge, Globalization, New Economy and the Environment. Business and Society Challenges for Sustainable Development *Proceedings of the 7th International Conference of the Russian Society for Ecological Economics* Saint Petersberg
LASP (2005) Legislative Assembly of St Petersburg, Zakonodatelnoje Sobranije St Peterburga,Zakon Sankt Peterburga o Generalnom Plane Sankt Peterburga I Granizah Zon Okhrany Objektov Kulturnogo Nasledija na Territorii Sankt-Peterburga Available online at: http://www.assembly.spb.ru/manage/page?tid=633200015&nd=8422495&nh=0&ssect=0

Leach MA (1997) A Systems Approach to Materials Flow in Sustainable Cities: A Case Study of Paper *Journal of Environmental Planning and Management* 40.6 705-723

Lindström M (2003) *Attitudes Towards Sustainable Development. Priorities, Responsibility, Empowerment* Lund, Lund University Press

London 21 (1998) Available online at: http://www.london21.org/

Lynch K (1960) *The Image of The City* Cambridge, MA, Technology and Harvard U Press

Medvedev VM and Aldasheva AA (2001) *Ekologicheskoe soznanije* Moscow, "Logos"

Milgram S (1974) *Obedience to Authority* New York, Harper & Row

Moiseev NN (2000) *Sudba Tsivilizacii*. Put' Razuma, Moscow, «A. Koshelev»

Muñiz I and Galindo A (2005) Urban Form and the Ecological Footprint of Commuting. The Case of Barcelona *Ecological Economics* 55.4 499-514

Nefedov VA (2002) *Landshaftnyj Dizajn I Ustojchivost' Sredy* St Petersburg, Poligrafist

Nijkamp P and Pepping G (1998) A Meta-Analytical Evaluation of Sustainable City Initiatives *Urban Studies* 35.9 1481-1500

Nijkamp P, Ouwersloot H and Rienstra SA (1997) Sustainable Urban Transport Systems: An Expert-Based Strategic Scenario Approach *Urban Studies* 34.4 693-712

Nijkamp P, Capello R and Pepping G (1999) *Sustainable Cities and Energy Policies* New York, Springer

Orr DW (2004) *The Nature of Design: Ecology, Culture, and Human Intention* New York, Oxford University Press

Proshansky HM (1978) The City and Self-Identity *Environment and Behavior* 10.2 147-169

Proshansky HM, Fabian AK and Kaminoff R (1983) Place-Identity: Physical World Socialization of the Self *Journal of Environmental Psychology* 3.1 57-83

Ravetz J (2000) *City Region 2020: Integrated Planning for a Sustainable Environment* London, Earthscan

REC (2000) Available online at: http://www.rec.org/REC/Programs/SustainableCities/What.html

Rees P (2002) *Urban Environment and Wildlife Law. A Manual for Sustainable Development* Oxford/Malden, Blackwell Science

Rees WE (2003) Understanding Urban Ecosystems: An Ecological Economics Perspective in Berkowitz AR, Nilon CH and Hollweg KS *Understanding Urban Ecosystems: A New Frontier for Science and Education* New York, Springer 115-136

Schmuck P and Schulz W (2002) *Psychology of Sustainable Development* Boston, Kluwer Academic

Shane AM and Graedel TE (2000) Urban Environmental Sustainability Metrics: A Provisional Set *Journal of Environmental Planning and Management* 43.5 643-663

Shmeleva IA (2006) *Psychologija Ekologicheskogo Soznanija*

Shmeleva IA and Shmelev SE (2007) *Sustainable Urban Development: Interdisciplinary Approach* St Petersburg, St Petersburg University Press

Shteinbach HE (2004) *Psychologija Zhiznennogo Prostrastva* St Petersburg, Rech

Swyngedouw E and Heynen N (2005) *In the Nature of Cities, Urban Political Ecology and the Politics of Urban Metabolism* London, Routledge

UN-HABITAT (2008) Available online at: http://www.unhabitat.ru/ru/about/history/

United Nations (1976) The Vancouver declaration on human settlements Available online at: http://www.unhabitat.org/downloads/docs/924_21239_The_Vancouver_Declaration.pdf

United Nations (1992) Agenda 21 Available online at: http://www.un.org/esa/sustdev/documents/agenda21/index.htm

United Nations (1996a) *Indicators of sustainable development: Framework and methodologies* New York, Commission on Sustainable Development

United Nations (1996b) Istanbul declaration on human settlements Available online at: http://www.unhabitat.org/content.asp?cid=2072&catid=10&typeid=25&subMenuId=0

United Nations (2002) World summit for sustainable development, Johannesburg Available online at: http://www.un.org/events/wssd/

United Nations (2005) UN population division of the Department of Economic and Social Affairs of the United Nations Secretariat, world population prospects: The 2004 revision and world urbanization prospects: The 2005 revision Available online at: http://esa.un.org/unup

United Nations (2007) *Indicators of sustainable development, guidelines and methodologies* (3rd edition) New York, United Nations

Vernadsky VI (1936) *Nauchnaya mysl kak planetarnoe javlenije*, Moscow, "Nauka"

Wackernagel M and Rees W (1996) *Our Ecological Footprint: Reducing Human Impact on the Earth* Gabriola Island, New Society Publishers

WCED (1987) *World Commission on Environment and Development, Our Common Future* New York, Oxford University Press

Wiek A and Binder C (2005) Solution Spaces for Decision Making – A Sustainability Assessment Tool for City-Regions *Environmental Impact Assessment Review* 25 589-608

Yagya V (2005) *Rassvet nad Peterburgom. Sankt-Peterburg v Mirovom Soobschestve* St. Petersburg, Evropejskij Dom

Chapter 11
Regional Waste Management: Multicriteria Modelling

Abstract This chapter is devoted to sustainable waste management. It summarises some recent work exploring the development of a multi-criteria optimisation tool for achieving sustainable solutions for municipal solid waste management systems (MSWMS). Aiming to provide a new methodological background for regional solid waste management modelling it takes into account spatial and temporal patterns of waste generation and processing, environmental as well as economic impacts of the system's development with a particular emphasis on public health and biodiversity.

Three different approaches to the spatial-temporal analysis of the MSWMS are used together, namely a life cycle inventory analysis, which helps to identify emission patterns within the MSWMS, a multi-criteria optimisation approach, which helps to find compromise solutions among environmentally and economically preferred options, and a geographic information systems approach, which provides a tool for identifying waste management facilities, transportation environmental and social impacts; it also gives an analysis of environmental impacts on valuable ecosystems. A Russian methodology for calculating environmental damage was used to weight the importance of different sub-territories covered by the system as well as simplifying the analysis of emissions from waste treatment plants. The approach provides a new perspective for the analysis of municipal solid waste management systems on the regional scale. The principal novelty of the proposed complex MSW strategic management model is an integration of the different types of data: geographical, environmental and economic, using relational database technology.

Simulations using the dataset for Gloucestershire illustrate the performance of a simplified version of the model. Simulations were undertaken to explore the potential effects on the waste management infrastructure of introducing the EU Landfill Directive. The chapter shows how methodological synthesis and a systems perspective give useful support to the decision-maker regarding potential development paths and trade-offs between the economic and environmental performance of a proposed waste management system.

Keywords Ecological–economic modelling • Waste management • Optimisation • Life cycle analysis • UK

DOI 10.1007/978-94-007-1972-9_11,

Strategic Waste Management Planning

Strategic decision-making for dealing with municipal solid waste is a problem currently exercising the minds of many local governments throughout the European Union (EU) and across the world. This chapter is devoted to ecological–economic modelling of strategic developments in Municipal Solid Waste (MSW) Management Systems at the regional level.

The waste management problem in the EU is characterised by increasing per capita production of waste materials (Figs. 11.1–11.3), the need for high levels of investment in physical infrastructure (incinerators, landfill sites, recycling facilities), institutional barriers (such as the long-term nature of contracts), a wide range of stakeholders and a dynamic policy arena (e.g. the Waste Electrical and Electronic Equipment and Landfill Directives are two instruments aimed at reducing the amounts of biodegradable and electronic waste being landfilled). The waste stream itself varies in composition over time and space with seasonal and longer term changes in the quantity and amounts of various materials and the market for 'recycled' materials is characterised by uncertain demand and fluctuating prices. Strategic decision-making for waste is a complex problem which appears to offer scope for

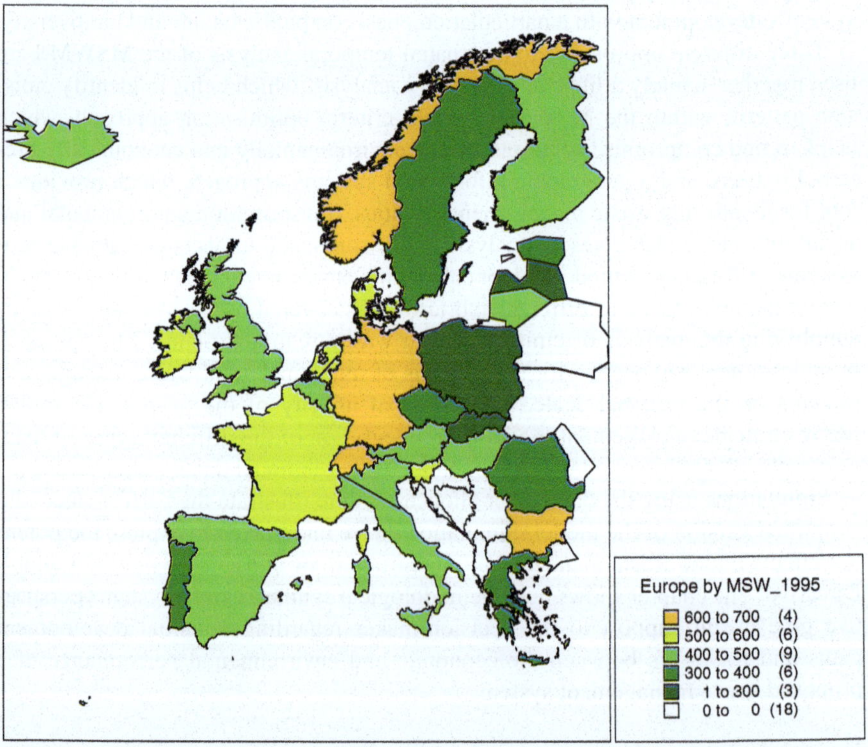

Fig. 11.1 Municipal solid waste generation per capita, EU, 1995

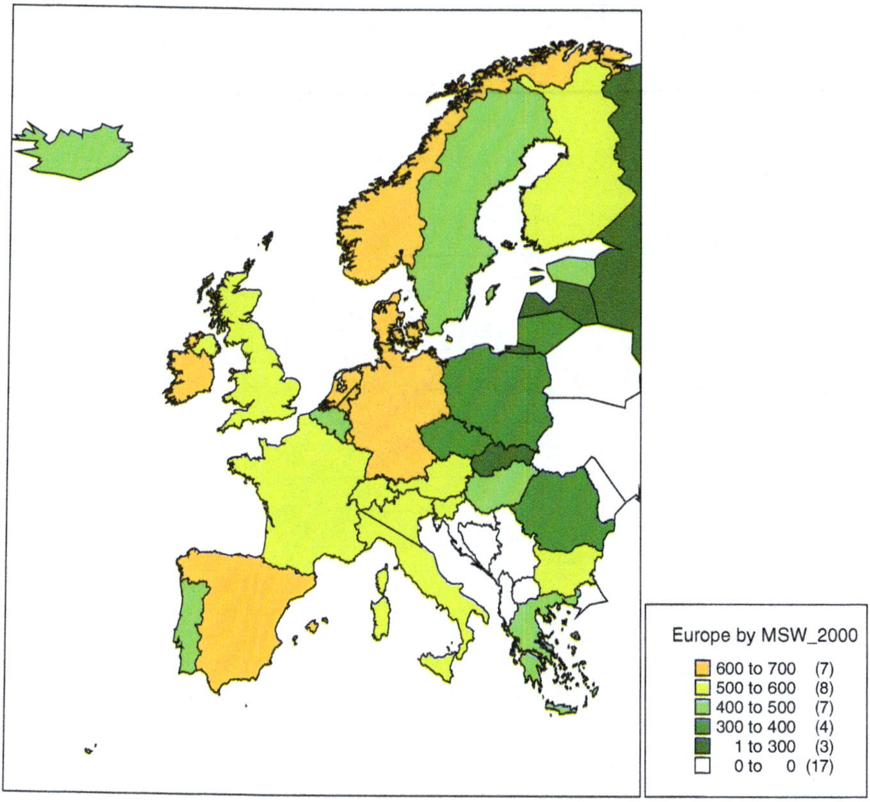

Fig. 11.2 Municipal solid waste generation per capita, EU, 2000

mathematical modelling procedures in order to find "optimal" solutions. Although standard modelling approaches are limited as the ideal solution looks very different depending where you are situated: from the household or local government point of view the best solution would be to eliminate the waste and remove the need for any waste service provision in the first place, while the view from the waste industry would be one of maximising the number of waste streams and quantities of waste over time to ensure survival of the industry. This paper considers whether ecological economic modelling approach has anything new to offer the policymaker.

Description of the Waste Management Problem

Ecological–economic modelling is an aid to strategic decision-making for waste management where there is nearly always strong local opposition to the siting of waste facilities, where alternative waste management approaches place heavy demands on the environment, and where future EU policy threatens to put the onus on producer responsibility and thus remove significant quantities of high value

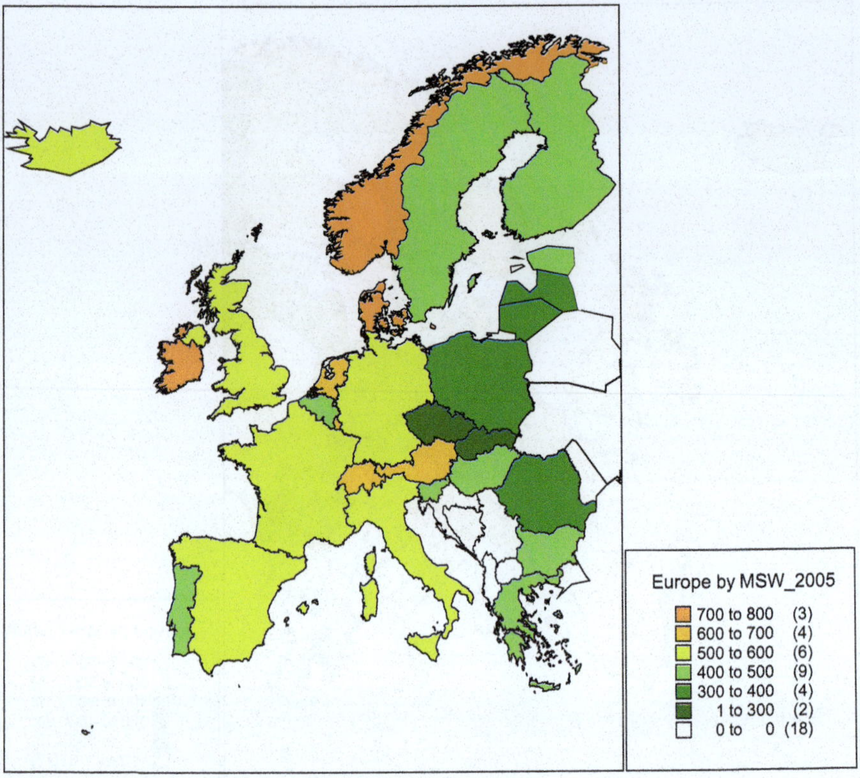

Fig. 11.3 Municipal solid waste generation per capita, EU, 2005

materials from the waste stream, and solutions are driven as much by local politics as by economic factors. Standard economic modelling approaches seeking the optimum or least cost solution fail as they cannot incorporate the wide range of factors which need to be included in a decision which must be based on achieving the Best Practicable Environmental Option (BPEO). Decision-makers do need assistance in making strategic choices which cause social and environmental impacts, and tie up large amounts of money and land for significant periods of time. The approach presented here is a first step in developing an ecological economic modelling approach which attempts to integrate life cycle inventory analysis, environmental impact assessment and economic appraisal within a geographic information system (GIS) framework. The aim is not to provide an "optimum" solution but to highlight to decision-makers the trade-offs inherent in investing in different mixes of waste management technology at a range of scales from the local to the regional. In other words, it can reveal, for a particular area or region, how waste management should be 'integrated' in order to achieve the BPEO solution.

The UK has made promising progress in this field (Fig. 11.4), but it is still lagging behind most Western European states in the way it treats its municipal waste. We can see that the share of MSW going to landfill in the UK has been steadily

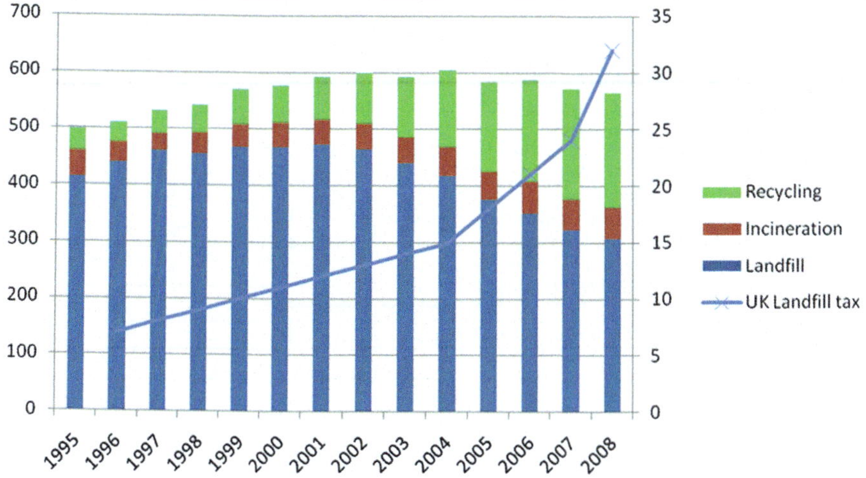

Fig. 11.4 Municipal solid waste treatment and landfill tax, UK (1995–2008)

decreasing from the year 2001, which happened to some degree due to increases in landfill tax, introduced in 1996.

The waste management problem has a complex nature with a range of important dimensions such as multiplicity of types of waste generated in the system, complex spatial pattern of waste generation, the necessity to transport waste long distances for processing, a variety of emissions from waste collection, transporting and treatment which affect the environment, and the almost unpredictable and localised character of the impact of these emissions on humans and ecosystems. And although there have been attempts to analyse regional waste management systems taking into account the environmental impacts of processes under study, most of them have not formed a holistic method for analysing all spatial, temporal and qualitative aspects of the problem. Therefore, the aim of this chapter is to provide a new methodological background developing regional municipal solid waste management modelling, taking into account spatio-temporal patterns of waste generation and processing, environmental as well as economic impacts of development of the system with a particular emphasis on public health and biodiversity.

This chapter takes the first steps in developing a model for municipal solid waste management system at the regional level. The paper analyses the post-consumption stages of the waste life cycle, namely collection, sorting, treatment and final disposal. The municipal solid waste management system under study is illustrated by Fig. 11.5, which shows the main material flows within the system. The figure reveals that the whole life cycle of materials entering and leaving the waste management system consists of several stages – raw materials extraction, processing, sale, consumption, finally becoming waste when they are discarded by consumers.

These materials in the waste stream then undergo collection, sorting (removal of recyclable materials) and treatment (which can be thermal or biological), with the final stage being disposal in landfill. The shaded areas in the diagram are the stages of the life cycle explicitly taken account of in this chapter.

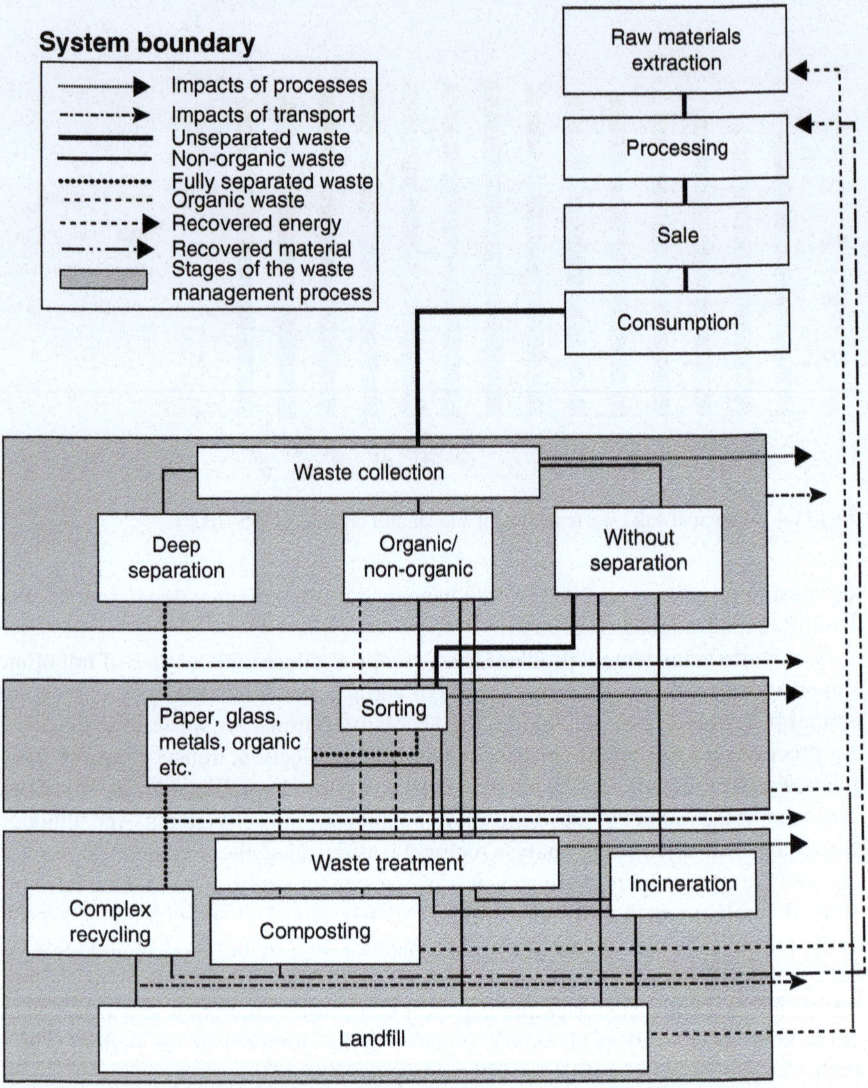

Fig. 11.5 The municipal solid waste management system: material flows

Approaches to Waste Management Modelling

There have been many attempts to analyse municipal solid waste management systems over the past decade. Economic as well as environmental and social aspects of their performance have been taken into account. Despite the large amount of research done, the application of the major methods employed does not provide a holistic picture of municipal solid waste management systems which can examine environmental impacts and the economic costs of siting, technological processes involved, transportation, impacts and their spatio-temporal distribution, or identify the parties affected (Fig. 11.5). The main spheres of research in the field of MSWMS in 1990s have been: analysis of waste generation determinants (Chen and Chang 2000, Daskalopoulos 1998, Hockett 1995), siting of waste management facilities (Chang and Wang 1997, Fredriksson 2000, Huang et al. 1995), the choice of waste treatment method (Dalemo 1998, Highfill 2001, Huhtala 1997), environmental impacts of different waste management technologies (Nixon 1997, Powell et al. 1996, Slater 2001), economic mechanism of managing MSWS (Fullerton 1998, Hong 1999, Jenkins 2000, Morris 1994, Palmer 1997), transportation of waste (Bhat 1996, Kulcar 1996), macroeconomics of recycling (Ferrer and Ayres 2000, Masui 2000, Nakamura 1999) and complex planning (Chang and Lin 1997, Chang and Wang 1997, Haastrup et al. 1998, Huang 1997). In the majority of this research, the focus has been on single aspects of the problem, for example, (Chang and Wang 1997) looked at management costs, air pollution and the recycling goals, but missed out water and soil pollution, noise, road congestion, employment and health impacts; Haastrup et al. (1998) concentrated on costs, air, water and soil pollution, road congestion, technological reliability, but did not cover noise, employment, health impacts or recycling goals.

A substantial amount of research on local aspects of municipal solid waste management modelling has been carried out using LCI methodology based on the recent models developed by White et al. (1999) and the Environment Agency's WISARD model. J.C. Powell (Powell et al. 1996), for example, compared environmental and social impacts of a kerbside collection scheme for recyclable household waste with a bring scheme, using life cycle assessments and economic valuation for assigning relative weights to these impacts, while Powell et al. (Powell 1998) explored alternative approaches to waste management for six district councils in Gloucestershire. Powell (2000) investigated the potential for using LCI analysis in local authority waste management decision making.

Many aspects of waste management systems performance were not integrated in a holistic model taking into account spatial distribution of environmental as well as economic impacts, nor were transportation, technological and siting issues analysed simultaneously. Unfortunately, almost all these models are of minimal use by decision-makers as they miss some of the key institutional dimensions of waste management as identified by Vigileos (2002), in particular the unequal social impacts of waste management, the nature of contracts drawn up between the industry and local authorities which are long-term, and sometimes require delivery of guaranteed amounts of waste, political pressures to recycle, barriers imposed by government regulations and the lack of communication between different participants in the waste management sector.

Although there are examples of environmental–economic analysis of municipal solid waste management systems on the regional level by Chang and Lin (1997), Haastrup (1998), and Chang and Wang (1996), many applications do not incorporate an integrated analysis of environmental impacts from all stages of the life cycle of municipal solid waste, spatial ecological–economic modelling of the

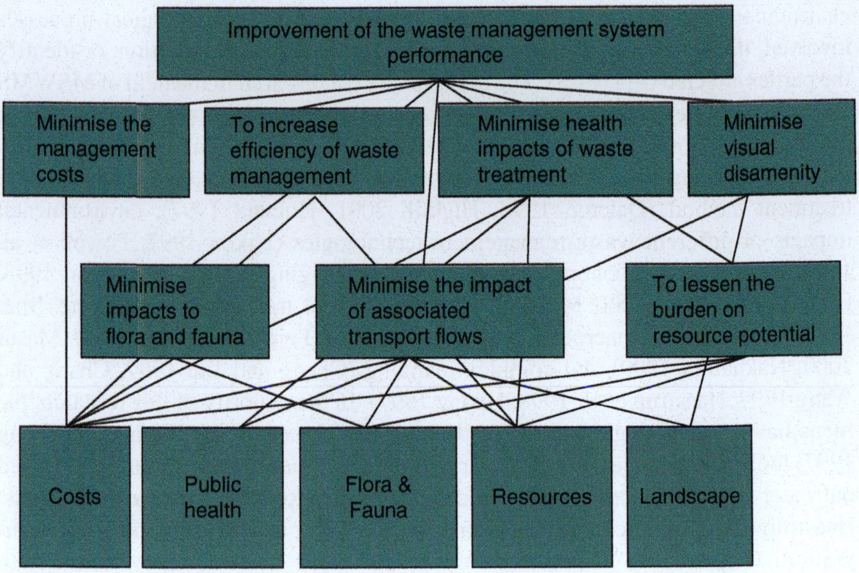

Fig. 11.6 Aspects of sustainable MSW management problem

distribution of impacts, non-substitutable treatment of environmental and economic characteristics of the development of the system, including non-monetary valuation of environmental damage (Fig. 11.6). What is missing is a technique for solving regional waste problems which inevitably have a large number of possible solutions due to variable population densities, incomes, multiple (actual and potential) locations for waste management infrastructure, protected landscape areas and high value ecological sites. There is thus an urgent need for improved methods for identifying BPEO solutions to waste management problems at the regional level. The range of potential development paths for a solid waste management system, for example, could include a large centralised regional facility, or a set of small localised ones, depending on the physical conditions, and this represents a situation of choice between multidimensional scenarios.

A Comparison of Methodological Approaches

Among the methods used for analysis of MSWMS during the past 10 years several should be mentioned here: input–output approach (Ferrer and Ayres 2000, Nakamura 1999), multiple regression analysis (Daskalopoulos 1998, Hockett 1995), life cycle analysis (Craighill and Powell 1996, Powell 1998, 2000, Song and Hyun 1999), operations research methods (Chang and Wang 1997, Chang et al. 1997), multi-criteria assessment (Hokkanen and Salminen 1997, Rogers and Bruen 1998,

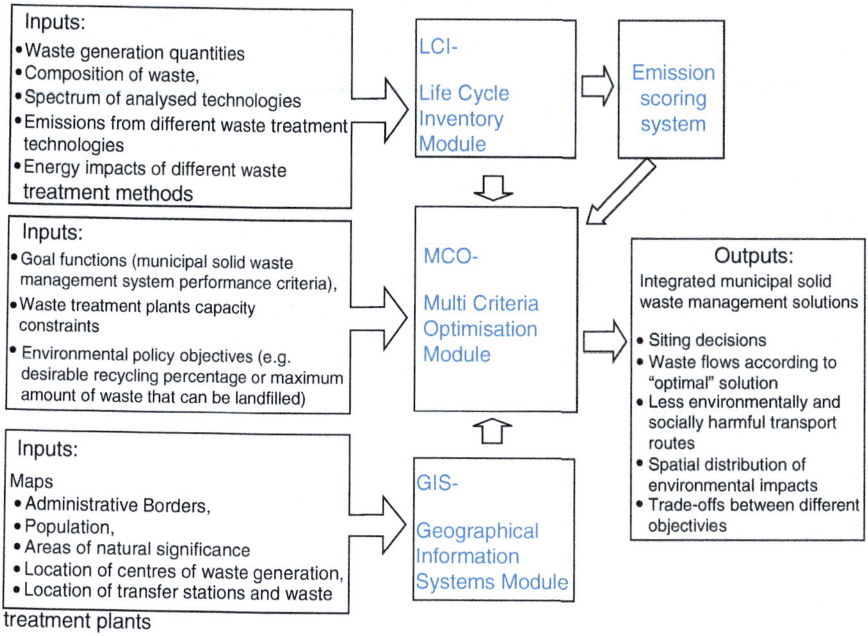

Fig. 11.7 Conceptual diagram of the modules of the decision support system

Salminen 1998) and expert systems (Barlishen and Baetz 1996, Haastrup et al. 1998). All of these methods have particular uses in specific areas and Table 26 below identifies their strengths and weaknesses.

The life cycle inventory approach (see Fig. 11.7) provides information on the spectrum and quantity of emissions from a given technological process and when it comes to comparing different scenarios, sophisticated methods of multi-criteria assessment (Munda and Romo 2001) may be applied. However, LCI methodology does not include any geographical or time dimension nor does it provide any estimate of the effect of the emissions inventoried. When used in isolation, it cannot identify the best solution (i.e. BPEO) of the waste management problem.

MCDA, optimisation, Delphi on the other hand allow for comparison between alternatives which need to be integrated with an approach which can analyse the waste management system itself.

Geographic Information technology is a powerful tool for analysing and exhibiting spatial data. However, the rating and scoring of several scenarios (which is done often in geospatial environmental impact assessment, EIA) is not sufficient for performing an integrated analysis of the development of the municipal solid waste management system.

It is necessary to perform a significant number of simulation experiments, changing different spatial siting patterns, processing capacities, and waste collection and sorting schemes to arrive at the decision space from which a selection can be made. All these approaches need to be underpinned by some impact assessment methodology. The one selected here is the Russian methodology for environmental damage calculation which was developed by Balatsky et al. (Vremennaja tipovaja metodika 1983, Vremennaja metodika 1999) and allows for taking into account the spatial dimension of environmental impacts in the form of coefficients of environmental value of the territories or regions (Vremennaja metodika 1999). At the same time, it reduces the dimension of the analysed vector of environmental characteristics of the given waste management system, which can be in turn divided into the negative effects of recycling, incineration and other waste treatment options, as well as negative effects on air, water and soil. Such reducing of the dimension simplifies the decision-making significantly and allows for dealing with only two dimensions of the waste management planning problem – environmental and economic.

The Russian methodology for estimating environmental damage uses coefficients of environmental harm, attributed to each type of emission into water and into air. These coefficients are developed from laboratory based biological research on animals (i.e. standard toxicological studies) and extrapolation of these effects on humans. This information is then integrated with another set of coefficients – coefficients of the environmental value of the territories or regions which are based on the ecosystem value of the major biomes, soils, water reserves, located in the territory of the given region.

In summary, we can say that LCI is good at modelling the waste system but is only a first stage in identifying environmental impacts as it concentrates on emissions to water, land and air, but does not provide any indication of the impact or significance of emissions locally. It needs to be integrated with other techniques such as EIA which can provide the impact analysis needed based on the siting of infrastructure, or movement of waste and with some optimisation procedure which can begin to deal with the issues of trading off economic costs and benefits against the social and environmental impacts of alternative waste management systems.

Thus, it is clear that what is required is a combination of several methods in order to perform the complex analysis of the potential development of the municipal solid waste system.

Several studies have already tried to combine some of these methods. During the 1990s, for example, GIS and EIA were combined into a geo-spatial EIA (Antunes et al. 2001, Patil et al. 2002), GIS and MCDA were combined by Dai et al. (2001), LCI and MCDA were integrated by Powell et al. (1996), and Munda and Romo (2001) and Powell et al. (1999) integrated a simple multi-criteria approach to examine environmental impacts from alternative waste management scenarios for the city of Bristol.

In summary, these studies are still limited and cannot be used to solve regional waste problems because they have not yet elaborated all the complex factors influencing waste management processes at the regional level – namely spatial distribution of waste generation, impact of transportation and processing of waste as well as the multidimensional character of these emissions, the time dimension in waste generation, the building of new or expansion of existing facilities and the spatial distribution of impacts of waste treatment processes on humans and on valuable ecosystems. This chapter draws on research (Shmelev 2003, Shmelev and Powell 2004) which focused on the integration of three different approaches to the spatio-temporal analysis of the MSWM problem, namely a life cycle inventory approach (LCI module), which helps to identify emission patterns within the MSWMS, a multi-criteria optimisation approach (MO module), which helps to find compromise solutions among the environmentally, economically and socially preferred options, and a geographic information systems approach (GIS module), which provides a base for siting waste management facilities, transportation, social impacts, and assessing environmental impacts on valuable ecosystems. A Russian approach to calculating environmental damage was utilised to weigh the importance of different sub-territories covered by the system. It is hoped that this approach will provide a new perspective for the analysis of municipal solid waste management systems.

Development of the Integrated Methodology

Based on an understanding of the weaknesses of the methods identified in Table 11.1 above, it was decided that by combining different methods a more useful tool might be developed for the development of strategic municipal solid waste management planning. The aim therefore was to develop an integrated technique which would give useful support to the decision maker regarding potential development paths and tradeoffs between the economic and the environmental performance of alternative waste systems.

Research carried out in Russia (St. Petersburg and the region) and the UK (Gloucestershire) concentrated on a complex analysis of the MSWMS, taking ecological, economic and social aspects of the management of municipal solid waste into account.

Due to software limitations, it was decided to limit the analysis of the municipal solid waste management system to the examination of five major components: i.e. the economic costs of running the system, public health, the state of the flora and fauna, saving of material resources and landscape quality. Four out of the five factors chosen to characterise the waste management system relate to the main goals of the EU Landfill Directive (European Council 1999) (reduction of adverse effects of landfill of waste on the environment, in particular on surface water, groundwater, soil, air and human health) and also correspond to the most relevant themes of the UN Sustainable Development Indicators: social (health), environmental (atmosphere, biodiversity) and economic (consumption and production patterns) (United Nations 2001). The fifth component, 'landscape quality', was selected to reflect the important role which landscape plays in local communities as stated in the European Landscape Convention, which identifies landscape as "a key element of individual and social well-being". Under Chap. 2 of the Convention, signatories agree to "integrate landscape into...regional and town planning policies...as well as in any other policies with possible direct or indirect impact on landscape" (Council of Europe 2000).

Table 11.1 Analytical tools for municipal solid waste decision making

Method	Strengths	Weaknesses
LCI – life cycle inventory	• Reflects a wide spectrum of emissions • Allows integration of environmental data with economic data • Flexible, allows easy comparison of different scenarios	• Only an inventory of emissions • No information on impacts to the recipients • No time or space related dimensions • Unable to make local/regional/global trade-offs
MCDA – multi-criteria decision analysis	• Allows comparison of multi-attribute or multi-objective scenarios • Flexibility in the choice of criteria • Allows integration of quantitative and qualitative data	• Problems with weight estimation • Limitations by comparing only a relative small number of alternatives, which could not represent the efficient set of solutions
Optimisation	• Gives the best solution from the feasible set • Permits solving of multi-objective problems by employing goal programming, compromise programming techniques, etc. • Allows the user to identify the efficient frontier of the solution space for subsequent decision-making	• The opportunity to solve large scale non-linear mixed integer problems limited by the existing algorithms • Certain assumptions about the relationships in the model have to be made
GIS – geo-information systems	• Reflects spatial patterns of the geographical distribution of actors, flows and sensitive areas • Allows the user to perform geographic analysis based on	• Does not have a time dimension • Requires integration with other techniques for performing comparative analysis of scenarios intersection, overlapping of different objects, etc. • The amount of output information is too high for decision-making
Environmental damage calculation methodology, Russia (1983, 1999)	• Allows for integration of many types of emissions into a single measure of environmental damage • Explicitly takes into account geographical peculiarities of the given territories	• No common and recognised measurement unit of environmental damage • No account taken of the receptors of polluting emissions

(continued)

Table 11.1 (continued)

Method	Strengths	Weaknesses
Delphi method	• Allows the user to analyse complex situations with uncertain information and/or lack of time/resources for decision-making using experts	• Subjectivism of estimates • Possibilities of unequal understanding the problem in question by the experts
Environmental impact assessment (EIA)	• Allows detailed examination of all the impacts from specific sites and technologies • Can combine economic, environmental and social information	• Very expensive in terms of time, resources, data demands • Necessary to combine with dispersion modelling • Very superficial types of studies • Focus is on the impacts and not the waste system itself
Pollution dispersion models	• Show detailed spatial distribution of emissions given the relief, climate and the characteristics of the source of emissions	• Substantial computational power is needed (esp. for multiple sources) • A very expensive tool • Difficult to analyse the impacts on the final recipients

These five components offered a relatively simple and straightforward means of analysing economic–environmental trade-offs. The monetary costs of operating a waste management system are of critical concern to local authorities, materials savings (i.e. recycling) are of national concern, and for strategic decision-making purposes both these elements need to be directly compared with the impacts on environmental and social factors (i.e. human health, environmental 'health' and landscape quality). The data on the selected components were also available and relatively easy to obtain.

Life cycle analysis using the model was integrated with a GIS and an optimisation technique. The LCA model allowed the researchers to examine a wide range of emissions from alternative waste management scenarios; the GIS allowed actual and proposed waste management sites, along with ecological sensitivity of the landscape to be mapped; the single criteria optimisation technique permits the possibility of deriving a unique solution of the problem.

The most difficult choice was that of the optimisation procedure. There are well known, rapid and reliable methods for solving linear problems, whereas it becomes more complex when the situation requires mixed-integer programming. There are ways of reducing the multidimensional problems to single-criterion ones, however, and several objectives may be taken into account simultaneously. The type of optimisation problem employed here (linear mixed integer programming problem) is complex and demands significant computational power and efficient algorithms, especially for real scale modelling. Constraints on resources and computational power led to a focus on a two-dimensional problem by examining single-criteria overall system cost minimisation with simultaneous calculation of an additional parameter (such as the environmental damage caused by the system performance). Although limited, this two-dimensional solution space nevertheless provides a useful starting point for understanding how useful such an integrated methodology might be.

The large sets of heterogeneous data used in the model (geographical, economic, environmental and social) are integrated using relational database technology. The database system consists of several interrelated tables representing different aspects of the problem under study (e.g. different types of waste analysed, spatially distributed waste generation centres, a range of waste treatment facilities, a multitude of emission types, etc.).

Modules within the Integrated Method

The GIS Module

The key elements of the GIS module are the digitised maps of the county of Gloucestershire, UK, obtained from a range of different sources. The maps are overlaid and allow graphical analysis of the location of the physical waste infrastructure and transport routes in relation to environmentally sensitive areas and centres of population density. The census ward was taken as a minimal geographical unit for population data.

The Impact Assessment Module

It should be noted that the methods of analysis and comparison of the emission inventory results within life cycle analysis is an area open to debate. In some cases, the list of emissions analysed numbers several hundred items. In order to deal with this vast amount of information in the current research, the methodology expressed in the Vremennaja metodika (1999) and Vremennaja tipovaja metodika (1983) was taken as an instrument for comparing scenarios with heterogeneous outputs. The list of substances taken into account in the analysis can be seen in Appendix 1. The toxicity coefficients database for all pollutants allows conversion of the wide spectrum of different substances into a unified index of environmental damage, which reduces the dimension of the problem substantially and simplifies the solution procedure.

The method is used here to provide the spatial dimension of environmental damage around waste treatment infrastructure sites in the form of the coefficients of the significance of the territories adjacent to the waste treatment plants. Such coefficients were derived by performing a series of operations on GIS maps. The dispersion of pollutants from the various waste treatment facilities was approximated by a 5-km radius circle around each of the sites. The coefficients of significance were derived based on the weighting of sensitive areas by a group of experts based at the University of Gloucestershire using a Delphi approach. Standard national designations of ecological and landscape importance were utilised by the experts: Sites of Special Scientific Interest (SSSI), National Nature Reserves (NNR), Special Areas of Conservation (SAC), Specially Protected Areas (SPA), RAMSAR sites and an indicator of population density. The average number of people living within a 5-km radius of the waste treatment plants was calculated using the average population density of neighbouring wards covered by the circle. The experts were asked to rate their perceptions of the relative sensitivity of the designated areas to the potential emissions from the waste management facilities on a 1–10 scale. This information was then integrated into a Randomised Preferences method (Hovanov 1996), which took the relational data derived from the experts to derive weighting factors. A Delphi approach was utilised because of the complexity and uncertainty over the impacts of the regional waste management system on the different aspects of the human and natural environments. The Delphi approach provided a quicker and cheaper alternative to more narrowly defined pollution dispersion modelling approaches. The Delphi approach has the added advantage of enabling localised priorities to be integrated into the significance measures, but the methodology employed needs to be transparent in order to understand the trade-offs generated. Thus, such an approach will provide variations in significance measures in different regions, related to population geography and protected areas.

Next overall indices of the importance of the territories around the waste treatment plants were obtained. First, all the 5-km radius circles around the waste treatment plants were overlayed onto maps of the different types of sensitive areas (including centres of population density).

Then the percentage of the intersection of each circle by each type of the sensitive area was multiplied by the importance factor for a given sensitive territory and it was summated over all six types of areas analysed according to the formula:

$$I_k = \sum_{j=1}^{J} F_j * \left(\frac{S(C_k \cap E_j)}{S(C_k)} \right)$$

where I_k is the importance score of the circle around kth waste treatment plant; F_j is an importance factor for the environmental sensitive territory type j, j = 1…J; S denotes area; C is an area of the circle around a waste treatment plant; E_j is a joint object consisting of the parts of environmentally sensitive areas falling within a given circle C_k.

The borders of each of the geographical objects are stored in the digital database with additional information such as the name of the object, areas and geographical coordinates. The centre points of the census wards are used to define the waste generation sites, and transport routes are considered here as the links connecting the centroids of the wards and the waste treatment plants.

The LCI Module

In the framework of the analysis carried out here, the life cycle analysis is bounded on the one side by the post-consumption generation of waste and on the other side by final disposal. It includes the analysis of a municipal solid waste stream comprising eight components – paper, glass, ferrous and non-ferrous metals, plastics (film), plastics (rigid), textile, organic and "other".

For each of the types of waste mentioned three basic treatment technologies are analysed: recycling, incineration with energy recovery and landfilling. The emissions to air, water and soil are analysed. In order to produce an integral index of environmental damage, the amounts of the polluting emissions are multiplied by the respective coefficients of environmental harm, according to Vremennaja metodika (1999) and Vremennaja tipovaja metodika (1983) as described above.

Optimisation Module

This module integrates the information on plant locations and distances between centres of population density and waste treatment plants from the GIS module and information on the level of emissions from each type of waste, collected, sorted and treated by each of the technologies from the LCI module. It is here that the choice of collection systems, sorting and treatment technologies, and the geographical distribution of aste management facilities are optimised over the time period of interest according to the total system cost minimisation criteria. The problem which is being analysed here belongs to the class of linear mixed-integer programming problems. LINGO optimisation software uses branch and bound methods to solve problems of this type. The information on the problem dimensions for the Gloucestershire case study is laid out in Table 11.2.

The initial problem is set in a single-criteria cost minimisation framework. The reason for this is that all the improvements in the environmental performance of waste management systems are bound by the budgets of the relative administrative units, and cost minimisation is still the dominant criteria for waste management system development. The environmental damage is calculated here as a by-product of the minimum costs scenario according to the formula:

$$ED = \sum_{k=1}^{K} I_k * \sum_{k=1}^{K} \gamma^l * E_k^l,$$

Table 11.2 Real problem dimensions for Gloucestershire

Set	Definition	Quantity of elements
J	Waste generation points	145
H	Waste types	9
I	Waste treatment centres	86
K	Waste treatment technologies	6
T	Periods of system functioning	20

The total amounts to 13,560,480 variables in the mixed integer programming model, including integer variables: 92,880; real variables: 13,467,600 and number of constraints: 13,591,278 (without trivial constraints, stating the non-negativity of the decision variables – 123,678)

where ED is total systems environmental damage, I_k is an importance score of the territory around k^{th} waste treatment plant, E_{kl} is the level of emissions of the lth type ($l = 1 \ldots L$) (from the LCI module) and γ_l is an environmental damage coefficient for the emission type l.

The final two-dimensional solution space in the form presented in Fig. 11.13 is obtained by performing a sensitivity analysis, which varies waste treatment capacity, relative to landfill space. It allows the decision-maker to analyse the given waste management system in terms of the trade-offs between environmental and economic objectives.

The simplified version of the model utilised here, based on the work of Baetz and Neebe (1994) was built using LINGO 7.0 optimisation software. The model only permitted examination of a reduced set of problem dimensions: three waste treatment technologies – recycling, incineration and landfilling were considered over 10 time periods with no consideration of the spatial dimension. The dimensions of the problem were determined by the constraints in the number of Boolean and continuous variables in the Demo version of LINGO 7.0. Due to resource constraints (software limitations), the model has not been realised to its full potential.

Two versions of the model were developed, which use Open Data Base Connectivity (ODBC) technology for transferring data from one software package to the other. The usefulness of the developed model is that it allows the user to change the initial data outside the model and then to "plug-in" the new datasets for subsequent solving. It is useful in the case of sensitivity analysis involving a large number of parameter changes.

It should be stressed here that solving this problem in real dimensions with standard tools would require handling vectors of model variables with 15,000,000 components. This will definitely require using more powerful database management systems (e.g. SQL Server) and the problem could be solved faster if its special structure could be taken into account.

The problems which can be analysed include the choice of waste processing technology (e.g. among landfilling, composting, waste-to-energy incineration, recycling), waste-management-facility-plant siting and the optimisation of the whole MSWMS performance using different goal functions.

The next section examines a simple application of the model to waste management in Gloucestershire.

Table 11.3 Composition of municipal solid waste in Gloucestershire, 1998/1999

Material	Gloucestershire average, %	National average, %
Fines	2.3	7
Ferrous	4.0	6
Glass	3.4	9
Green	11.3	21
Putrescibles	34.4	
Misc.com.	5.8	8
Misc.non-com.	0.5	2.20
Non-ferrous	1.0	2
Paper and card	20.8	32
Plastic film	4.9	5
Rigid plastics	7.6	6
Textiles	3.9	2

Gloucestershire figures do not include recycled waste

Case Study of Gloucestershire

Gloucestershire lies in the west of England (South West Region), has a total area of 2,618,000 km^2 and a population of 574,000 (2001). Gloucestershire comprises six local authorities: Cheltenham Borough, Cotswolds District, Gloucester City, Forest of Dean District, Stroud District and Tewksbury Borough.

Average number of people in households is 2.41. The average disposable income per person per year is £10,073 (1999, data for the South West Region), and annual waste generated range from a low of 280 to a high of 432 kg of municipal solid waste per person per year in Gloucestershire. The annual recycling rate in 1998/1999 ranged from 6% in Gloucester to 19% in Cotswolds District. The dominant municipal solid waste treatment method is landfilling (82% in the South West Region of the UK).

The average composition of municipal solid waste in Gloucestershire is presented in Table 11.3 and Fig. 11.8 illustrates location of waste facilities.

Simulations using the dataset for Gloucestershire were performed on the simplified version of the model. Eight thousand one hundred and forty simulations were undertaken (see Fig. 11.8), where the waste treatment capacities for recycling, incineration and landfilling were changed. This could illustrate, for example, the possible consequences of introducing the EU Landfill Directive in the county, which may result in fewer landfills and increased recycling capacity, with consequent impacts on transport routes and costs across the county. The objective of the directive is to prevent or reduce as far as possible negative effects on the environment from the landfilling of waste, by introducing stringent technical requirements for waste and landfill, and by reducing the quantity of biodegradable material going to landfill. The scenarios examined here show the potential effects of reductions in available landfill space as a result of the Directive and explore the impact of increased tipping fees and recycling subsidies on the environmental and economic performance of the system.

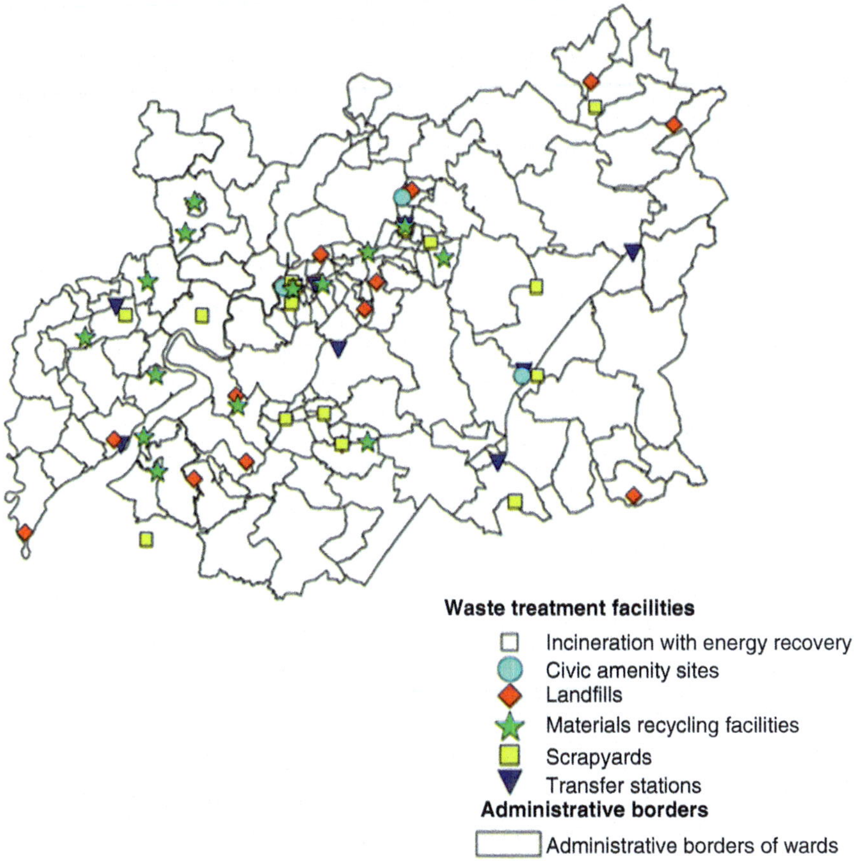

Fig. 11.8 Regional waste management system in Gloucestershire, UK

Figure 11.13 illustrates the combinations of minimal costs and corresponding environmental damage for the whole range of scenarios examined. All the combinations of potential environmental damage and economic costs are given here under equal economic conditions. Only the landfill and waste treatment capacities were changed in this analysis.

Results of the Simulation Experiments

The results of a series of simulation experiments are depicted in Figs. 11.9–11.13. The study of the developed model of the regional waste management system was conducted along the following main lines: it was decided to study the sensitivity of the model first to the changes in technological parameters of the available capacity of

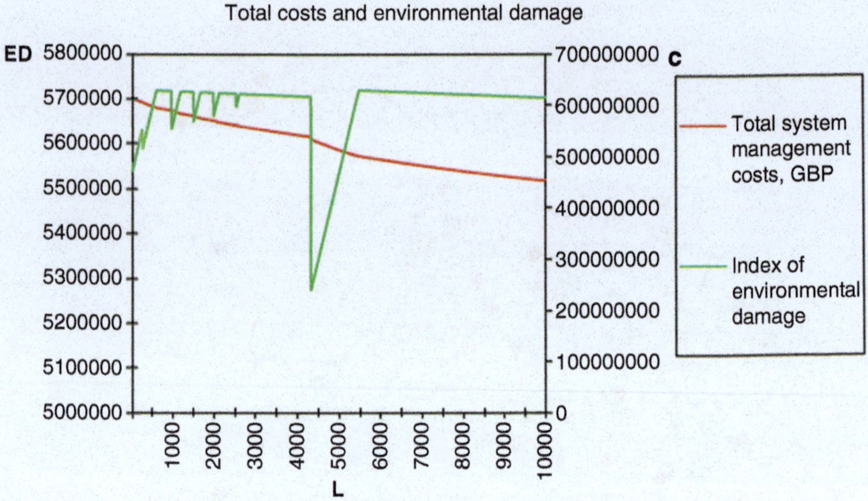

Fig. 11.9 Scenario 3, RE = 200; W = 200; LL = 5,000

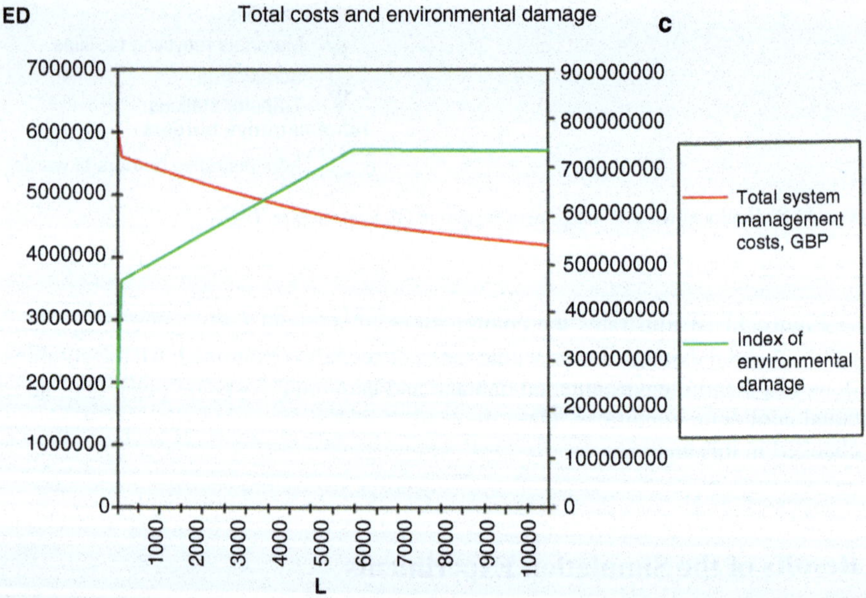

Fig. 11.10 Scenario 4, RE = 600; W = 200; LL = 0

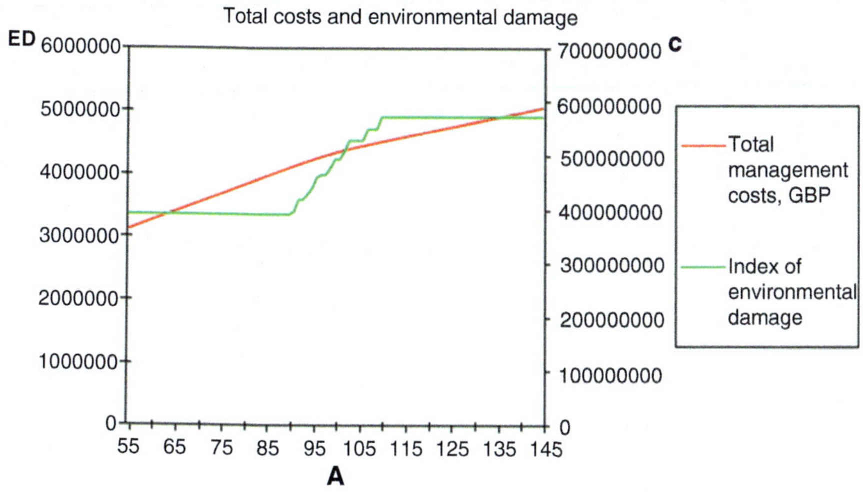

Fig. 11.11 Scenario 6, RE=600; W=200; L=5,000; LL=0, illustrating changes in A

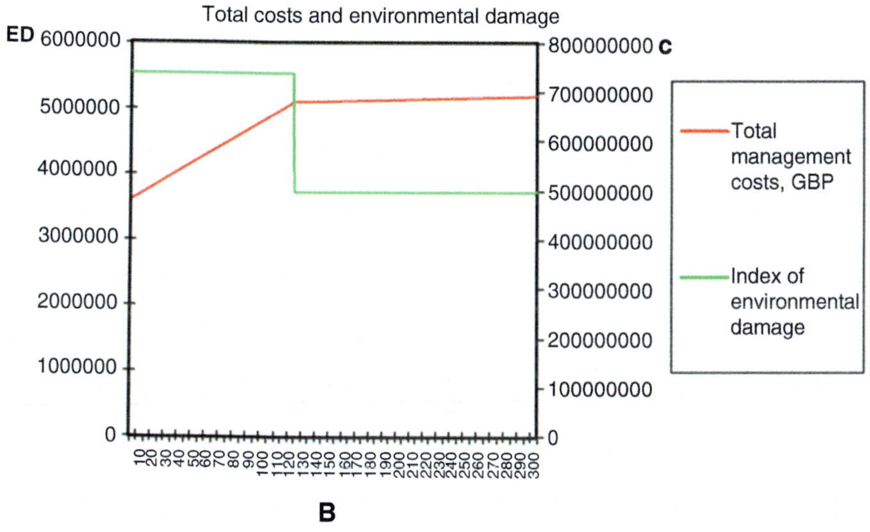

Fig. 11.12 Scenario 7, RE=600; W=200; L=5,000; LL=0, illustrating changes in B

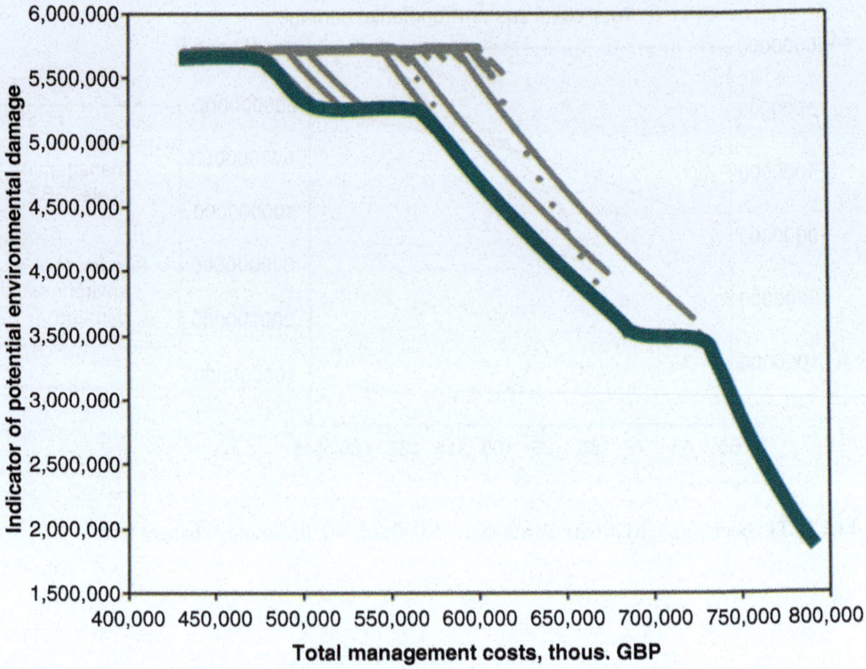

Fig. 11.13 Two-dimensional solution space

Table 11.4 Parameters changed in sensitivity analysis

Set of simulation experiments	LL	RE	W	L	Changed parameter	The interval of change
1	0	200	200		L	10:10,000
2	1,000	200	200		L	10:10,000
3	5,000	200	200		L	10:10,000
4	0	600	200		L	10:10,000
5	0	400	400		L	10:10,000
6	0	600	200	5,000	A	55:145
7	0	600	200	5,000	B	10:300

L available capacity of the existing landfill, thousands m^3, *LL* available capacity of the additional landfill, thousands m^3, *RE* recycling capacity, thousands of tonnes per year, *W* incinerating capacity, thousands of tonnes per year

existing landfill under different combinations with other technological parameters being fixed, and later to changes in price parameters – the cost of recycling of a tonne of waste and costs of collection and transporting waste to landfill. The combinations of the system parameters used in sensitivity analysis are shown in Table 11.4.

ED – environmental damage denotes the index of environmental damage and C – costs denotes total management costs in British pounds.

In the first Scenario, recycling and incineration capacity was limited to 200,000 tonnes/year; there was no opportunity to open an additional landfill site.

With parameter L decreasing at first against a background of considerable growth in costs, slow growth in environmental damage takes place, caused by intensive use of incineration as an alternative to decreasing landfill capacity; then with decreasing L<5,500 the substitutional transition to recycling part of the waste takes place, causing a considerable decrease in environmental damage by a factor of 1.08.

In Scenario 2, there is an option of opening a small additional landfill with the capacity of 1,000,000 m^3. The first local minimum of environmental damage is found at L=5,500. Such a sharp decrease in environmental damage is caused by the growth in recycling, instead of harmful use of landfill sites; the subsequent growth in damage is caused by the opening of an additional landfill site in the 9th period; and the rapid decrease in environmental damage starting at L=4,500 can be explained by the ever increasing rate of recycling. At the same time the costs are naturally starting to grow at a faster rate.

In scenario 3 (Fig. 11.9), the rapid decrease in environmental damage as L approaches the value of 5,500 is caused by the growth of the proportion of recycling; the rapid growth in environmental damage after L<4,500 is caused by the opening of a new landfill site for 5,000,000 m^3 in the 6th period, with simultaneous decrease in the share of recycling and incineration. After that, the proportion of waste being incinerated increases. The following local minima can be explained by the shift of the moment of opening an additional landfill site in the 5th, 4th period and so on.

The tendency for environmental damage and minimal management costs to change in scenario 4 (Fig. 11.10) might be divided into three different stages – 10,000>L>5,500, 5,500>L>100, L<100. In the first stage, gradual growth in the proportion of waste being incinerated takes place, this causes slower growth in environmental damage and costs; decreasing environmental damage and costs growing at the faster rate in the second stage are caused by the growth in the proportion of waste undergoing complex recycling at L<100; when landfilling capacity for placing even incineration residue becomes critical, the shift towards recycling on a larger scale takes place.

In the 5th scenario, everything develops similarly to the 4th; however, due to the larger planned incineration capacity and smaller recycling capacity, the shift to the second stage of intensive recycling takes place later, at about L=750, and to the third, earlier, around L=200.

The sensitivity of the solution of the problem to changes in price parameters is illustrated in Figs. 11.11 and 11.12.

Analysing the changes in environmental damage caused by the decreasing price of complex recycling of a tonne of waste (parameter A, recycling costs, Fig. 11.12), we come to a conclusion about regarding the lack of changes in environmental damage with parameter A being reduced from 145 to 110. Then, the sharp decrease in environmental damage – by more than a factor of 1.7 with the subsequent decrease in A to 80, and again, at the interval, [55…80] environmental damage is at a lower level than in the first case, but is nevertheless stable.

Changes in the parameter B – costs of collection and transportation of waste to the landfill site in Landfill 1 could suggest the optimal level for transport costs set up in the interests of environmental protection (under conditions of legal waste discharges by the companies and municipalities). The results of the simulations experiments which may be seen in Fig. 11.12 show that, as transport costs increase up to a certain level (in our case B = 120) and given that the law is observed, transporting waste to landfill may become less desirable than recycling.

The main result of the work – two-dimensional solution space, which is an integration of the results of sets of simulation experiments 1–5 (Table 11.4), shows that, by increasing total system management costs by a factor of 1.82, it is possible to diminish total environmental damage by a factor of 2.99.

The shape of the thick curve representing the set of non-dominated solutions (solutions which are equal or at least not worse than the rest) depicts the peculiarities of the complex problem of the development of a waste management system, giving the decision-maker the range of options he or she can choose from and thereby helping him trade-off economic versus environmental aspects of the development of the system in question. We are definitely not proposing "the best solution" or BPEO to the decision-maker, but providing him or her with freedom of informed choice, however, hard it may be to make a choice.

This latter aspect appears in the realm of pure political decision making.

Discussion

The results presented here illustrate an application of a simplified ecological–economic model of a municipal solid waste management system. Full development of the model would facilitate the solution of more complex problems involving real decisions of siting, choice of treatment technology, collection and sorting method. Certain weaknesses remain in the approach taken here, primarily software limitations and probably lack of pollution dispersion modelling.

The main strength of the model is that it allows the decision- maker to analyse the ecological–economic trade-offs in the development of the municipal solid waste management system. It examines possible strategies for the development of the system, taking into account different siting options and choice of waste treatment technologies; it allows preliminary investment planning and explicitly takes account of the spatial dimension of environmental impacts on public health and valuable ecosystems.

In the life cycle analysis performed here, the boundaries are defined by post-consumption solid waste generation through to the moment of final disposal. If the boundaries were altered to include elements related to the production of waste processing equipment, the transportation fuel life cycle, analysis of materials and products the solid waste was derived from, results could change significantly.

The model presented in this chapter might be developed further to take into account the real dimensions of the problem, such as transportation of waste, improved pollution dispersion models and the introduction of hyperbolic discounting (Daly and Farley 2004). If we take into account the origins of waste, and work on material flows accounting of products entering the system in the first place then, with programming improvement, a full scale decision-support tool for strategic regional waste management might be created. The next steps are to apply more powerful software, possibly to integrate pollution dispersion models for all sources of pollution and to analyse more rigorously the chains of impacts. It might be valuable to integrate the analysis of the environmental impacts of transportation, taking into account noise and congestion impacts.

Models of this type might then be expanded and applied at a regional level in the EU, in order to provide improved information on the tradeoffs to be made in what are inherently difficult political problems.

Appendix 1 The List of Emission Coefficients

Sector of the ecosystem	Emission type	Recycling	Incineration	Landfilling	Damage coefficients
Air	Particulates	0.00327	0.00002	0	2.7
Air	CO	0.00228	0.0004	3.125E–06	0.4
Air	CO_2	0	1.1293	0.2209825	0.4
Air	CH_4	0	0	0.098215	0.7
Air	NOx	0.00231	0.0016	0	16.5
Air	N_2O	0.000053	0	0	30
Air	SOx	0.003947	0.0003	0	20
Air	HCl	0.0000033	0.0001	1.625E–05	20
Air	HF	5E–09	0	3.25E–06	500
Air	H_2S	0.000012	0	0.00005	500
Air	HC	0.001692	0.0001	0.0005	20
Air	Chlor. HC	0	0.0001	8.75E–06	50
Air	Dioxins/furans	0	5E–13	0	50,000
Air	NH_3	0.0000004	0	0	28.5
Air	As	0	0.0000025	0	500
Air	Cd	0	0.0000005	1.4E–09	500
Air	Cr	0	0.0000063	1.65E–10	1670
Air	Cu	0	0.0000063	0	500
Air	Pb	0	0.0000063	1.275E–09	5000
Air	Hg	3E–09	0.0000005	1.025E–11	5000
Air	Ni	0	0.00000025	0	500
Air	Zn	0	0.00000063	1.875E–08	500
Air	Landfill gas (250 nm^3/t) generation (t/t)	0	0	250	0
Water	BOD	0.00239	0	0.0004751	5
Water	COD	0.02084	0	0.0004751	2
Water	Sus. sol.	0	0	0.000015	0.15
Water	TOG	0.000004	0	0.0000003	50
Water	AOX	0.0000025	0	0.0000003	1000
Water	Chlor. HCs	0	0	1.545E–07	0
Water	Dioxins/furans	0	0	4.8E–14	0

(continued)

(continued)

Sector of the ecosystem	Emission type	Recycling	Incineration	Landfilling	Damage coefficients
Water	Phenol	0	0	5.7E–08	0
Water	NH_4	4.47E–07	0	0.0000315	1
Water	Tot. metals	0	0	1.442E–05	0
Water	As	0	0	2.1E–09	90
Water	Cd	0	0	2.1E–09	250
Water	Cr	0	0	9E–09	550
Water	Cu	0	0	8.1E–09	550
Water	Fe	0	0	1.425E–05	1
Water	Pb	0	0	9.45E–09	11
Water	Hg	0	0	9E–11	15,000
Water	Ni	0	0	2.55E–08	90
Water	Zn	0	0	1.02E–07	90
Water	Cl	0.000011	0	0.0000885	550
Water	F	9.7E–07	0	5.85E–08	550
Water	NO_3	0	0	0	0.2
Water	S^-	0.000006	0	0	550

Appendix 2 Types of Environmentally Sensitive Areas Taken into Account by the Model

AONB (Areas of Outstanding Natural Beauty) – the areas protected by the Government of the UK since 1949 "National Parks and Access to the Countryside Act". The main goal of the designation AONB is preservation of the natural beauty of the landscape, and recreational use is not a major goal here and is permitted only to the extent that such use is in accordance with the preservation of natural beauty and the needs of agriculture, forestry and other spheres of regional development as well as the economic and social interest of local communities. Such areas number 41 in 2002 – they cover approximately 15% of the territory of England and Wales.

SSSI (Sites of Special Scientific Interest) – the land designated as such according to the 1981 "Wildlife and Countryside Act" (UK) (as amended).

NNR (National Nature Reserves) – lands designated according to the "National Parks and the Access to the Countryside Act" of 1949 (UK).

SAC (Special Areas of Conservation) – lands, whose status is drawn in the EC Directive 92/43/EEC on Conservation of the natural environments, wild fauna and flora. The data acquired have a status "candidate".

SPA (Special Protection Areas) – lands, classified according to the EC Directive 79/409 on the preservation of wild birds. The data acquired has the status "classified".

RAMSAR (unique wetland complexes) – the land, which has a status of the Wetlands of International Importance according to the Ramsar convention. The Convention on Wetlands, signed in Ramsar, Iran, in 1971, is an intergovernmental treaty, which provides a framework for national action and international cooperation for the conservation and wise use of wetlands and their resources. There are presently 138 Contracting Parties to the Convention, with 1364.30 wetland sites, totalling 119.6 million hectares, designated for inclusion in the Ramsar List of Wetlands of International Importance.

Appendix 3 Data Requirements

The dynamic spatial ecological–economic model of the MSWMS built here links different types of data: GIS data sets, environmental impact information, economic information, specific waste related information, time information.

The required GIS data sets include:

- County, district and ward boundaries;
- General purpose layers: rivers and waterways, motorways, urbanised areas;
- Population density within wards;
- Areas of ecological significance: Sites of Special Scientific Interest (SSI), National Nature Reserves (NNR), Special Areas of Conservation (SAC), Special Protection Areas (SPA);
- Sites of existing and proposed waste management facilities;
- Distances between the points in question (between waste treatment plants and centroids of the chosen population areas), other characteristics of transport routes.

The environmental impact information needed will consist of:

- Emission coefficients of waste treatment by different technologies (recycling, RDF, landfilling, etc.), taking into account the analysed types of waste (paper, glass, etc.) and the list of substances of interest;
- Emission coefficients of using different types of fuel for transporting waste;
- Coefficients of environmental harm from different substances emitted into air and water according to Russian environmental damage estimation methodology;
- Expert weighting of relative importance of the environmentally sensitive areas examined with respect to placing waste treatment plants near them.

Economic information comprises:

- Costs of processing different types of waste by various technologies;
- Investment costs for building new waste processing plants;
- Transportation costs;
- Prices of recycled materials and energy derived from waste.

Specific waste related information:

- Types of waste under consideration;
- Respective technologies used for processing each of the types of waste;

- Waste composition in various districts;
- Sorting and collection information.

Time related information:

- Timescale of the model (number of periods under consideration, length of periods);
- Impacts which could differ over time (e.g. gaseous emissions from landfill). Time factor in economic decisions (discount factor).

References

Antunes P, Santos R and Jordao L (2001) The Application of Geographical Information Systems to Determine Environmental Impact Significance *Environmental Impact Assessment Review* 21.6 511-535

Barlishen KD and Baetz BW (1996) Development of a Decision Support System for Municipal Solid Waste Management Systems Planning *Waste Management & Research* 14.1 71-86

Baetz BW and Neebe AW (1994) A planning model for the development of waste material recycling programmes. *Journal of Operations Research Society* 45.12 1374-1384

Bhat VN (1996) A Model for the Optimal Allocation of Trucks for Solid Waste Management *Waste Management and Research* 14.1 87-96

Chang N-B and Lin WT (1997) Economic Evaluation of a Regionalization Program for Solid Waste Management in a Metropolitan Region *Journal of Environmental Management* 51.3 241-274

Chang N-B and Wang SF (1996) Solid Waste Management System Analysis by Multiobjective Mixed Integer Programming Model *Journal of Environmental Management* 48.1 17-43

Chang N-B and Wang SF (1997) A Fuzzy Goal Programming Approach for the Optimal Planning of Metropolitan Solid Waste Management Systems *European Journal of Operational Research* 99.2 303-321

Chang N-B and Wang SF (1997) Integrated Analysis of Recycling and Incineration Programs by Goal Programming Techniques *Waste Management and Research* 15.2 121-136

Chang N-B, Chen YL and Wang SF (1997) A Fuzzy Interval Multiobjective Mixed Integer Programming Approach for the Optimal Planning of Solid Waste management Systems *Fuzzy Sets and Systems* 89.1 35-60

Chen HW and Chang N-B (2000) Prediction Analysis of Solid Waste Generation Based on Grey Fuzzy Dynamic Modeling *Resources, Conservation and Recycling* 29.1–2 1-18

Council of Europe (2000) European Landscape Convention. http://conventions.coe.int/Treaty/en/Treaties/Html/176.htm

Craighill AL and Powell JC (1996) Lifecycle Assessment and Economic Evaluation of Recycling: A Case Study *Resources, Conservation and Recycling* 17.2 75-96

Dai FC, Lee CF and Zhang XH (2001) GIS-based Geo-environmental Evaluation for Urban Land-Use Planning: A Case Study *Engineering Geology* 61.4 257-271

Dalemo M (1998) Effects of Including Nitrogen Emissions from Soil in Environmental Systems Analysis of Waste Management Strategies *Resources, Conservation and Recycling* 24.3–4 363-381

Daly H and Farley J (2004) *Ecological Economics*. Principles and Applications. Island Press

Daskalopoulos E (1998) Municipal Solid Waste: A Prediction Methodology for the Generation Rate and Composition in the European Union Countries and the United States of America *Resources, Conservation and Recycling* 24.2 155-166

European Council (1999) Council Directive 1999/31/EEC of 26 April 1999 on the Landfill of Waste. http://europa.eu.int/comm/environment/waste/landfill_index.htm

Ferrer G and Ayres RU (2000) The Impact of Remanufacturing in the Economy *Ecological Economics* 32.3 413-429

Fredriksson PG (2000) The Siting of Hazardous Waste Facilities in Federal Systems *Environmental and Resource Economics* 15.1 75-87

Fullerton D (1998) Policies for Green Design *Journal of Environmental Economics and Management* 36.2 131-148

Haastrup P (1998) A Decision Support System for Urban Waste Management *European Journal of Operational Research* 109.2 330-341

Haastrup P et al (1998) A Decision Support System for Urban Waste Management *European Journal of Operational Research* 109.2 330-341

Highfill J (2001) Landfilling Versus "Backstop" Recycling When Income Is Growing *Environmental and Resource Economics* 19.1 37-52

Hockett D (1995) Determinants of Per Capita Municipal Solid Waste Generation in the Southeastern United States *Journal of Environmental Management* 45.3 205-217

Hokkanen J and Salminen P (1997) Choosing a Solid Waste Management System Using Multicriteria Decision Analysis *European Journal of Operational Research* 98.1 19-36

Hong S (1999) The Effects of Unit Pricing System upon Household Solid Waste Management: The Korean Experience *Journal of Environmental Management* 57.1 1-10

Hovanov NV (1996) Analis i sintez pokazatelej pri informazionnom defizite. St. Petersburg, SPb State Univ

Huang GH (1997) Capacity Planning for an Integrated Waste Management System Under Uncertainty: A North American Case Study *Waste Management and Research* 15.5 523-546

Huang GH, Baetz BW and Patry GG (1995) Grey Integer Programming: An Application to Waste Management Planning Under Uncertainty *European Journal of Operational Research* 83.3 594-620

Huhtala A (1997) A Post-consumer Waste Management Model for Determining Optimal Levels of Recycling and Landfilling *Environmental and Resource Economics* 10.3 301-314

Jenkins RR (2000) The Determinants of Household Recycling: A Material Specific Analysis of Recycling Program Features and Unit Pricing *Resources for the Future, DP 99-41-REV*

Kulcar T (1996) Optimizing Solid Waste Collection in Brussels *European Journal of Operational Research* 90.1 71-77

Masui T (2000) Analysis of Recycling Activities Using Multi-sectoral Economic Model with Material Flow *European Journal of Operational Research* 122.2 405-415

Morris GE (1994) The Economics of Household Solid Waste Generation and Disposal *Journal of Environmental Economics and Management* 26.3 215-234

Munda G and Romo M (2001) Combining Life Cycle Assessment and Multicriteria Evaluation: Comparing Waste Management Options in Spain *in* Spash C and McNally S *Managing Pollution and Environmental Toxicology* Cheltenham, Edward Elgar

Nakamura S (1999) An Interindustry Approach to Analyzing Economic and Environmental Effects of the Recycling of Waste *Ecological Economics* 28.1 133-145

Nixon WB (1997) An Empirical Approach to the Performance Assessment of Solid Waste Landfills *Waste Management and Research* 15.6 607-626

Palmer K (1997) The Cost of Reducing Municipal Solid Waste *Journal of Environmental Economics and Management* 33.2 128-150

Patil AA, Anachhatre AP and Tripati NK (2002) Comparison of Conventional and Geo-spatial EIA: A Shrimp Farming Case Study *Environmental Impact Assessment Review* 22.4 361-375

Powell J (1998) Using Life Cycle Inventory Analysis in the Development of a Waste Management Strategy for Gloucestershire, UK *Environmental and Waste Management* 1.4 221-233

Powell J (2000) The Potential for Using Life Cycle Inventory Analysis in Local Authority Waste Management Decision Making *Journal of Environmental Planning and Management* 43.3 351-367

Powell JC et al (1996) A Lifecycle Assessment and Economic Valuation of Recycling *Journal of Environmental Planning and Management* 39.1 97-112

Powell J et al (1999) *Life Cycle Inventory Analysis of Alternative Waste ManagementOptions for Bristol City Council: Summary Report* Cheltenham, Environmental Management Research Group, University of Gloucestershire, Cheltenham

Rogers M and Bruen M (1998) A New System for Weighting Environmental Criteria for Use Within ELECTRE III *European Journal of Operational Research* 107.3 552-563

Salminen P (1998) Comparing Multicriteria Methods in the Context of Environmental Problems *European Journal of Operational Research* 104.3 485-496

Slater RA (2001) Composting Municipal Waste in the UK: Some Lessons from Europe *Resources, Conservation and Recycling* 32.3–4 359-374

Song H-S and Hyun JC (1999) A Study on the Comparison of the Various Waste Management Scenarios for PET Bottles Using the Life-Cycle Assessment (LCA) Methodology *Resources, Conservation and Recycling* 27.3 267-284

Shmelev SE (2003) Ekologo-ekonomicheskoe modelirovanie regionalnych system upravlenia otchodami (Ecological–economic modelling of the regional waste management systems), Saint Petersburg State University, Russia — PhD thesis, 218 pp. (in Russian)

Shmelev S and Powell J (2004) Ecological–economic modeling of regional waste management systems. Proceedings of the 8th Biennial Scientific Conference *Challenging Boundaries: Economics, Ecology and Governance*. International Society for Ecological Economics, Montréal, Canada (July 11–14)

United Nations (2001) Indicators for Sustainable Development, CSD Theme Indicator Framework. http://www.un.org/esa/sustdev/natinfo/indicators/isdms2001/table_4.htm

Vigileos G (2002) Analysis of Institutional Structures for Sustainable Solid Waste Management for the South West of England, PhD thesis University of Gloucestershire, Cheltenham

Vremennaja metodika (1999) Vremennaja metodika opredelenijapredotvraschennogo ekologicheskogo uscherba Goskomekologia RF, Moscow, 1999

Vremennaja tipovaja metodika (1983) Vremennaja tipovaja metodikaopredelenija ekonomicheskoj effektivnosti osuschestvlenijaprirodoochrannych meroprijatij i ozenki ekonomicheskogo uscherba, prichinyaemogo narodnomukhozyajstvu zagryazneniem okruzhajushej sredy 1983 Moscow

White PR, Franke M and Hinde P (1999) *Integrated Solid Waste Management: A Lifecycle Inventory* London, 2009, Wiley-Blackwell; 2 edition, Oxford

Chapter 12
Business and Sustainable Development: CSR in Practice

Abstract This chapter focuses on the role of business in the quest for sustainable development. It addresses the concept of corporate social responsibility and outlines the history of sustainability reporting, placing emphasis on the Global Reporting Initiative. Central to the chapter is the concept of "stakeholder" as opposed to the "shareholder" of the profit maximising tradition of the past. The dynamics of CSR publication is shown alongside the full list of CSR reporting criteria. A new way of assessing the sustainability performance of companies is suggested and experience of development of CSR traditions in China, Japan, Germany, UK and France is reviewed. Cross-cultural differences in CSR discourse in the USA, UK and Germany are presented.

Keywords CSR • Corporate sustainability • GRI • Assessment • Discourse

Corporate Sustainability

Corporate sustainability has become a buzz-word in the past decade and a considerable amount of literature has been devoted to it in most recent years: (Dunphy 2003, Hand and Charity Finance Directors' Group (Great Britain) 2009, Henriques 2004, Steger 2004, Van Tulder and Van der Zwart 2006, Verbeke 2009, Werther 2011).

Conceptual articles on the new model of corporate social responsibility appeared as early as 1970s: (Carroll 1974, 1979), the more detailed discussion emerged in the 1990s: (Carroll 1991, 1999, Ulhoi 1995) followed by many others: (Azapagic 2003, Dyllick 2002, Figge and Hahn 2004, Miles et al. 2009, Stubbs and Cocklin 2008, Taneja et al. 2011, Welford 2002). Although originally some attention has been given to monetary assessment of environmental damages, which the author of this volume doesn't quite share (Atkinson 2000), more recently diverse indicator sets (Callens and Tyteca 1999, Wang and Lin 2007) and multi-stakeholder approaches become more popular (Angus-Leppan et al. 2010, Clifton and Amran 2011, O'Connor and Spangenberg 2008, Welford et al. 2008). There has been a steady

S.E. Shmelev, *Ecological Economics: Sustainability in Practice*,
DOI 10.1007/978-94-007-1972-9_12,

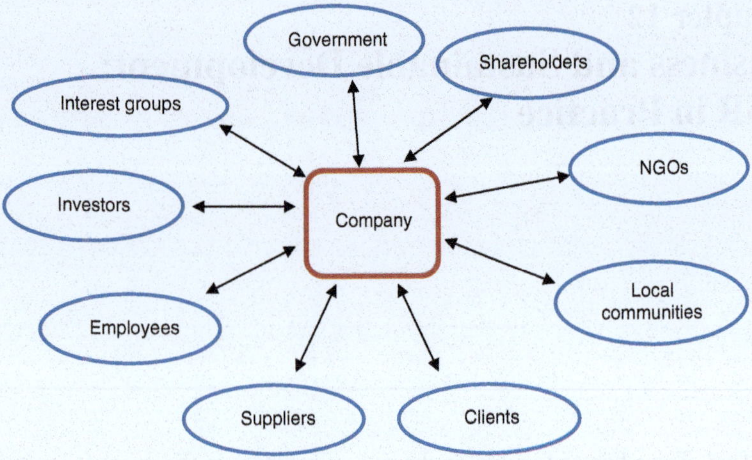

Fig. 12.1 Corporate stakeholders

interest in the systemic evolution of the Global Reporting Initiative (Brown et al. 2009a, b, Isaksson and Steimle 2009, Line et al. 2002) as well as experience of applying corporate sustainability principles in various regions of the world: Canada (Nitkin and Brooks 1998), US and the EU (Tschopp 2005), the UK (Idowu and Towler 2004), Germany (Gamerschlag et al. 2010), France (Delbard 2008), Latin America (Tschopp 2005), Sweden (Hedberg and von Malmborg 2003), Switzerland (Daub 2007), Brazil (Duarte 2010).

The main philosophy behind Corporate Social Responsibility or, as it is frequently referred to, Corporate Citizenship, Corporate Responsibility, Corporate Social Performance, Corporate Accountability, Sustainability, and Triple Bottom Line has been to include additional stakeholders (Fig. 12.1) in the corporate management framework. In other words, the management paradigm has shifted from "maximising profits for shareholders" to "creating value for a society at large", with the latter represented by the employees, clients, suppliers, local communities, investors, NGOs, government, and various interest groups (Table 12.1). Leading international organisations have developed strategies for the improvement of corporate performance in the field of sustainability, with UN Global Compact and its Ten Principles leading the way (Box 12.1).

Global Reporting Initiative

One such programme designed to influence corporate performance was The Global Reporting Initiative started in 1997–1998. The main idea behind this initiative was to create a new disclosure framework on sustainability at the corporate level so that companies might have an opportunity to show not only

Table 12.1 Factors driving enterprises to release CSR

Rank	Factor
	Company image
	Supporting government policy
	Leader's consciousness
	Public opinion pressure
	Demands from suppliers
	Investor pressure
	Demands from industry standards
	Abiding by laws and regulations
	Demands of capital markets
	Demands of innovation
	Mass consciousness
	Consumer pressure
	Power of NGOs
	Local community impact

Box 12.1 UN Global Compact. Ten Principles

Human Rights

- Principle 1: Businesses should support and respect the protection of internationally proclaimed human rights; and
- Principle 2: make sure that they are not complicit in human rights abuses.

Labour

- Principle 3: Businesses should uphold the freedom of association and the effective recognition of the right to collective bargaining;
- Principle 4: the elimination of all forms of forced and compulsory labour;
- Principle 5: the effective abolition of child labour; and
- Principle 6: the elimination of discrimination in respect of employment and occupation.

Environment

- Principle 7: Businesses should support a precautionary approach to environmental challenges;
- Principle 8: undertake initiatives to promote greater environmental responsibility; and
- Principle 9: encourage the development and diffusion of environmentally friendly technologies.

Anti-Corruption

- Principle 10: Businesses should work against corruption in all its forms, including extortion and bribery.

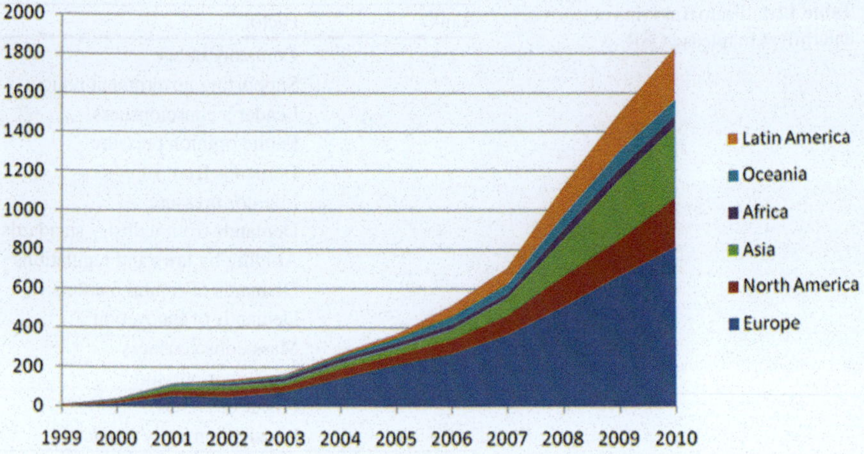

Fig. 12.2 Corporate sustainability reports according to GRI, 1999–2010

Table 12.2 GRI reporting dynamics (1999–2006)

GRI Reporting	
Year	Number of organisations
1999	20
2000	50
2001	80
2002	150
2003	325
2004	500
2005	750
2006	850+

their profits, assets and various financial ratios, but also their performance on the sustainability front. UNEP joined as a partner in 1999 and draft sustainability reporting guidelines were issued. The GRI's first Sustainability Reporting Guidelines were issued in 2000. The second generation of guidelines ("G2") was released in 2002. The third generation of the GRI guidelines ("G3") was produced in 2005.

Sustainability reports are usually complied by a company's management team internally or externally, with assurance provided by consultancies like CERES, SustainAbility, SGS, Corporate Citizenship, Ernst&Young, Deloitte, Det Norske Veritas, Two Tomorrows, CSR Network, Just Assurance, PriceWaterhouseCoopers, ERM Certification and Verification Services, Gerling Consulting Group GmbH, KPMG Sustainability B.V., Denkstatt GmnH.

According to the GRI website, there were over 1,800 companies, which produced a Corporate Sustainability Report following GRI standards in 2010. Exploring the evolution in CSR reporting patterns (Fig. 12.2 and Table 12.2) one may note the importance of European companies in this process, occupying around 50% of the list every year, as well as two important tendencies: the growth of the proportion of Asian and Latin American companies, particularly noticeable after 2005.

Corporate Sustainability Indicators

Sustainable Development reporting now follows the Global Reporting Initiative (GRI) guidelines (G3 edition) promoted by the UN Global Compact. The guidelines recommend the use of certain principles when compiling corporate sustainability reports. In order to define report content, principles of materiality, stakeholder inclusiveness, sustainability context and completeness are applied. In order to ensure report quality the principles of balance, comparability, accuracy, timeliness, clarity and reliability are used. G3 comes with a recommended set of 79 carefully selected indicators, which are grouped in the following categories:

- Economic (EC1 to EC9);
- Environmental (EN1 to EN30), subdivided into
  - Materials (EN1 to EN2);
  - Energy (EN3 to EN7);
  - Water (EN8 to EN10);
  - Biodiversity (EN11 to EN15);
  - Aspect: Emissions, Effluents and Waste (EN16 to EN25);
  - Aspect: Products and Services (EN26 to EN27);
  - Aspect: Compliance (EN28);
  - Aspect: Transport (EN29);
  - Aspect: Overall (EN30);
- Social, subdivided into:
  - Labour Practices & Decent Work (LA1 to LA14);
  - Human Rights (HR1 to HR9);
  - Society (SO1 to SO8);
  - Product Responsibility (PR1 to PR9).

Each group of indicators is divided in turn into Core, which are compulsory for disclosure and Additional, which are optional (Table 12.3).

The following strategy might prove useful for the analysis of corporate sustainability: sustainability reports of Company X might be assessed from the point of view of the latest GRI G3 reporting standards. Performance on each criteria might be seen as an element of a multicriteria decision matrix. Such an assessment might be performed for one company over time or for a whole sector in a comparative manner. On the basis of the results one can identify the areas of improvement for a company's overall sustainability performance, as well as its relative competitive

Table 12.3 GRI indicators of sustainable business performance

Economic			
Economic performance	Core	EC1	Direct economic value generated and distributed, including revenues, operating costs, employee compensation, donations and other community investments, retained earnings, and payments to capital providers and governments
Economic performance	Core	EC2	Financial implications and other risks and opportunities for the organisation's activities due to climate change
Economic performance	Core	EC3	Coverage of the organisation's defined benefit plan obligations
Economic performance	Core	EC4	Significant financial assistance received from government
Market presence	Add	EC5	Range of ratios of standard entry level wage compared to local minimum wage at significant locations of operation
Market presence	Core	EC6	Policy, practices, and proportion of spending on locally-based suppliers at significant locations of operation
Market presence	Core	EC7	Procedures for local hiring and proportion of senior management hired from the local community at significant locations of operation
Indirect economic impacts	Core	EC8	Development and impact of infrastructure investments and services provided primarily for public benefit through commercial, in-kind, or pro bono engagement
Indirect economic impacts	Add	EC9	Understanding and describing significant indirect economic impacts, including the extent of impacts
Environment			
Materials	Core	EN1	Materials used by weight or volume
Materials	Core	EN2	Percentage of materials used that are recycled input materials
Energy	Core	EN3	Direct energy consumption by primary energy source
Energy	Core	EN4	Indirect energy consumption by primary source
Energy	Add	EN5	Energy saved due to conservation and efficiency improvements
Energy	Add	EN6	Initiatives to provide energy-efficient or renewable energy-based products and services, and reductions in energy requirements as a result of these initiatives
Energy	Add	EN7	Initiatives to reduce indirect energy consumption and reductions achieved
Water	Core	EN8	Total water withdrawal by source
Water	Add	EN9	Water sources significantly affected by withdrawal of water

Water	Add	EN10	Percentage and total volume of water recycled and reused
Biodiversity	Core	EN11	Location and size of land owned, leased, managed in, or adjacent to, protected areas and areas of high biodiversity value outside protected areas
Biodiversity	Core	EN12	Description of significant impacts of activities, products, and services on biodiversity in protected areas and areas of high biodiversity value outside protected areas
Biodiversity	Add	EN13	Habitats protected or restored
Biodiversity	Add	EN14	Strategies, current actions, and future plans for managing impacts on biodiversity
Biodiversity	Add	EN15	Number of IUCN Red List species and national conservation list species with habitats in areas affected by operations, by level of extinction risk
Emissions, effluents, and waste	Core	EN16	Total direct and indirect greenhouse gas emissions by weight
Emissions, effluents, and waste	Core	EN17	Other relevant indirect greenhouse gas emissions by weight
Emissions, effluents, and waste	Add	EN18	Initiatives to reduce greenhouse gas emissions and reductions achieved
Emissions, effluents, and waste	Core	EN19	Emissions of ozone-depleting substances by weight
Emissions, effluents, and waste	Core	EN20	NO_x, SO_x, and other significant air emissions by type and weight
Emissions, effluents, and waste	Core	EN21	Total water discharge by quality and destination
Emissions, effluents, and waste	Core	EN22	Total weight of waste by type and disposal method
Emissions, effluents, and waste	Core	EN23	Total number and volume of significant spills
Emissions, effluents, and waste	Add	EN24	Weight of transported, imported, exported, or treated waste deemed hazardous under the terms of the Basel Convention Annex I, II, III, and VIII, and percentage of transported waste shipped internationally
Emissions, effluents, and waste	Add	EN25	Identity, size, protected status, and biodiversity value of water bodies and related habitats significantly affected by the reporting organisation's discharges of water and runoff
Products and services	Core	EN26	Initiatives to mitigate environmental impacts of products and services, and extent of impact mitigation
Products and services	Core	EN27	Percentage of products sold and their packaging materials that are reclaimed by category
Compliance	Core	EN28	Monetary value of significant fines and total number of non-monetary sanctions for non-compliance with environmental laws and regulations
Transport	Add	EN29	Significant environmental impacts of transporting products and other goods and materials used for the organisation's operations, and transporting members of the workforce
Overall	Add	EN30	Total environmental protection expenditures and investments by type

(continued)

Table 12.3 (continued)

Society			
Community	Core	SO1	Nature, scope, and effectiveness of any programmes and practices that assess and manage the impacts of operations on communities, including entering, operating, and exiting
Corruption	Core	SO 2	Percentage and total number of business units analysed for risks related to corruption
Corruption	Core	SO 3	Percentage of employees trained in organisation's anti-corruption policies and procedures
Corruption	Core	SO 4	Actions taken in response to incidents of corruption
Public policy	Core	SO 5	Public policy positions and participation in public policy development and lobbying
Public policy	Add	SO6	Total value of financial and in-kind contributions to political parties, politicians, and related institutions by country
Anti-competitive behaviour	Add	SO7	Total number of legal actions for anti-competitive behaviour, anti-trust, and monopoly practices and their outcomes
Compliance	Core	SO8	Monetary value of significant fines and total number of non-monetary sanctions for non-compliance with laws and regulations
Labour practices and decent work			
Employment	Core	LA1	Total workforce by employment type, employment contract, and region
Employment	Core	LA2	Total number and rate of employee turnover by age group, gender, and region
Employment	Add	LA3	Benefits provided to full-time employees that are not provided to temporary or part-time employees, by major operations
Labour/management relations	Core	LA4	Percentage of employees covered by collective bargaining agreements
Labour/management relations	Core	LA5	Minimum notice period(s) regarding significant operational changes, including whether it is specified in collective agreements
Occupational health and safety	Add	LA6	Percentage of total workforce represented in formal joint management-worker health and safety committees that help monitor and advise on occupational health and safety programmes
Occupational health and safety	Core	LA7	Rates of injury, occupational diseases, lost days, and absenteeism, and total number of work-related fatalities by region
Occupational health and safety	Core	LA8	Education, training, counselling, prevention, and risk-control programmes in place to assist workforce members, their families, or community members regarding serious diseases
Occupational health and safety	Add	LA9	Health and safety topics covered in formal agreements with trade unions. Health and safety topics covered in formal agreements with trade unions

Training and education	Core	LA10	Average hours of training per year per employee by employee category
Training and education	Add	LA11	Programmes for skills management and lifelong learning that support the continued employability of employees and assist them in managing career endings
Training and education	Add	LA12	Percentage of employees receiving regular performance and career development reviews
Diversity and equal opportunity	Core	LA13	Composition of governance bodies and breakdown of employees per category according to gender, age group, minority group membership, and other indicators of diversity
Diversity and equal opportunity	Core	LA14	Ratio of basic salary of men to women by employee category
Human rights			
Investment and procurement practices	Core	HR1	Percentage and total number of significant investment agreements that include human rights clauses or that have undergone human rights screening
Investment and procurement practices	Core	HR2	Percentage of significant suppliers and contractors that have undergone screening on human rights and actions taken
Investment and procurement practices	Add	HR3	Total hours of employee training on policies and procedures concerning aspects of human rights that are relevant to operations, including the percentage of employees trained
Non-discrimination	Core	HR4	Total number of incidents of discrimination and actions taken
Freedom of association and collective bargaining	Core	HR5	Operations identified in which the right to exercise freedom of association and collective bargaining may be at significant risk, and actions taken to support these rights
Child labour	Core	HR6	Operations identified as having significant risk for incidents of child labour, and measures taken to contribute to the elimination of child labour
Forced and compulsory labour	Core	HR7	Operations identified as having significant risk for incidents of forced or compulsory labour, and measures taken to contribute to the elimination of forced or compulsory labour
Security practices	Add	HR8	Percentage of security personnel trained in the organisation's policies or procedures concerning aspects of human rights that are relevant to operations
Indigenous rights	Add	HR9	Total number of incidents of violations involving rights of indigenous people and actions taken

(continued)

Table 12.3 (continued)

Product responsibility			
Customer health and safety	Core	PR1	Life cycle stages in which health and safety impacts of products and services are assessed for improvement, and percentage of significant products and services categories subject to such procedures
Customer health and safety	Add	PR2	Total number of incidents of non-compliance with regulations and voluntary codes concerning health and safety impacts of products and services, by type of outcomes
Product and service labelling	Core	PR3	Type of product and service information required by procedures and percentage of significant products and services subject to such information requirements
Product and service labelling	Add	PR4	Total number of incidents of non-compliance with regulations and voluntary codes concerning product and service information and labelling, by type of outcomes
Product and service labelling	Add	PR5	Practices related to customer satisfaction, including results of surveys measuring customer satisfaction
Marketing communications	Core	PR6	Programmes for adherence to laws, standards, and voluntary codes related to marketing communications, including advertising, promotion, and sponsorship
Marketing communications	Add	PR7	Total number of incidents of non-compliance with regulations and voluntary codes concerning marketing communications, including advertising, promotion, and sponsorship, by type of outcomes
Customer privacy	Add	PR8	Total number of substantiated complaints regarding breaches of customer privacy and losses of customer data
Compliance	Core	PR9	Monetary value of significant fines for non-compliance with laws and regulations concerning the provision and use of products and services

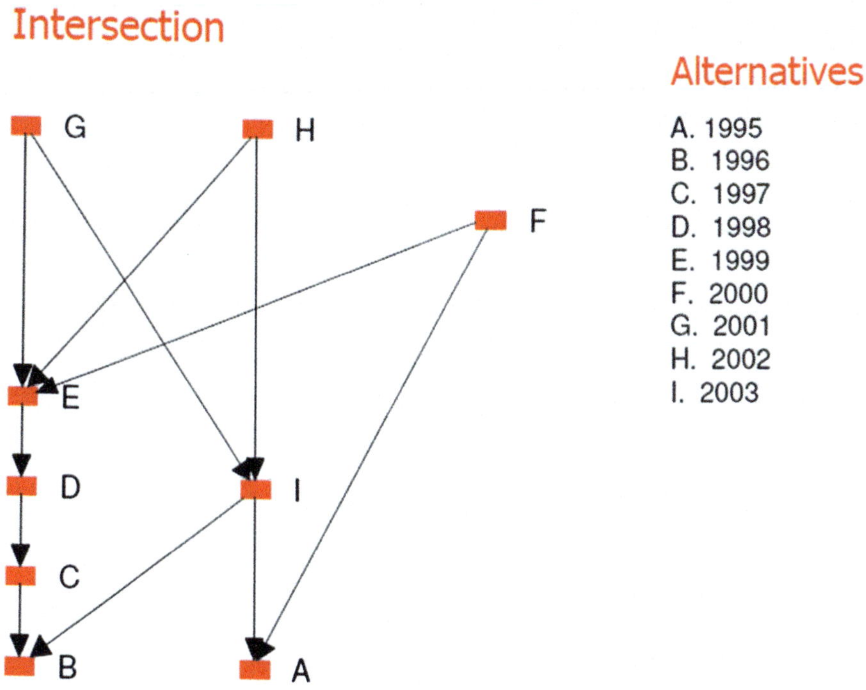

Fig. 12.3 Sustainability assessment chart using multiple criteria

position within the sector. Identification of gaps in available information will also help improve the quality of sustainability reporting.

The results will inform the management of a general multidimensional trend towards sustainability at the corporate level, taking all sustainability criteria into account simultaneously. This methodology was pioneered by the author and applied to various systems for improved sustainability analysis. An example of such an assessment is given in Fig. 12.3.

Figure 12.3, which represents the web of domination relationships among various years of evolution of a Company X, might be interpreted in the following way: each letter denotes a particular year of the Company's performance, A being 1995 and I being 2003. If there is a domination relationship between various years, it will be indicated by an arrow. As can be seen from the chart hypothetical Company X was most successful along the path towards sustainability in the years 2001 (G), 2002 (H) and 2000 (F), but was not so successful in the years 1995 (A) and 1996 (B). This allows us to reach a conclusion on the sustainability progress of Company X, giving complex and detailed advice. All the GRI recommended indicators or a smaller subset may be taken into account when making an assessment. Alternative methods capable of dealing with a small set of indicators or a very large set should be considered here.

On the basis of such an analysis, a complex of concise recommendations might be offered, addressing each area of sustainability performance in detail and offering practical steps for achieving further improvement on the path towards creating real sustainability value.

Cross-Country Comparisons

Undoubtedly it will be interesting to explore the extent to which CSR has been developed in various countries and the cross-cultural differences which can be indentified in the factors influencing the uptake of the CSR practices.

China (Xinyu et al. 2010) has been one of the most interesting cases in corporate sustainability reporting. The first ever CSR report in China was issued in 1999 by Shell and was followed by CNPC, Ford Motors, Baogang Steel, Ping An Insurance, Toshiba China and Jiangxi Mobile. Seven CSR reports were published in China in 2005, and 18 in 2006, reaching a total greater than 80 in 2008. For companies working in China, two major motivations for issuing CSR reports have been shown to be the most important: economic reasons and reasons of competitiveness in the case of multinationals working in China; and government regulation in the case of Chinese companies.

The case of Japanese companies is also quite special. The launch of CSR in Japan is usually associated with the year 2003 (Fukukawa and Teramoto 2009). In Japan, CSR is understood to describe those corporate principles or policies, which have long been influencing corporate activities, principles such as "to put utmost priority on respecting human dignity, safety and legal compliance" or "to contribute to society via our business". Sustainability is understood as a long-term pursuit of the company, as well as a means to support the successful continuation of its business. Most Japanese managers taking part in the survey referred to the word "globalisation" as one of the reasons for their company to adopt CSR management practices. Interestingly, the subject of human rights the managers found difficult.

In Germany (Gamerschlag et al. 2010) it was shown that on average, CSR disclosure is positively associated with company visibility, CSR disclosure is positively associated with profitability, CSR reports are more detailed for companies involved in the "heavier" or "polluting industries": automobile, chemicals, energy, utilities, construction; also, the presence of a company on the New York Stock Exchange has been shown to be related to the quality of its CSR reports.

The situation in the UK might be characterised by the presence of very large companies which embraced the CSR ideas and government support of the CSR process manifested in the creation of the CSR minister in 2000. The UK government in the 1970s passed many Acts of Parliament relevant to CSR (Idowu and Towler 2004) such as the Equal Pay Act 1970, Health and Safety at Work Act 1974, Sex Discrimination Act 1975, Race Relations Act 1976. The recommendations of the EU's Fifth Action Programme on the Environment summarised in the report "Towards Sustainability" (1992) contributed to the interest in CSR. The EC's 1993 Environmental Management and Audit Scheme (EMAS) encouraged companies to

Fig. 12.4 Concept map for US CSR reports (Chen and Bouvain 2009)

disclose relevant environmental information and initiate eco-auditing. Many companies have produced full-scale CSR reports (among them several banks which went totally bankrupt during the recession of 2007–2009); some have added a few pages of CSR information in their general reports. The results of the recent study show a weak positive relationship between the depth of CSR disclosure and the Earnings per Share for respected companies.

In France, the disclosure of the social and environmental impacts of a company's activities if it is registered on the Paris stock market became obligatory under the New Law on Economic Regulation (Blasco and Zølner 2010, Delbard 2008). Three hundred and seventy three French companies participate in the UN Global Compact and France ranks Number 4 in the world on the global report list with external assurance.

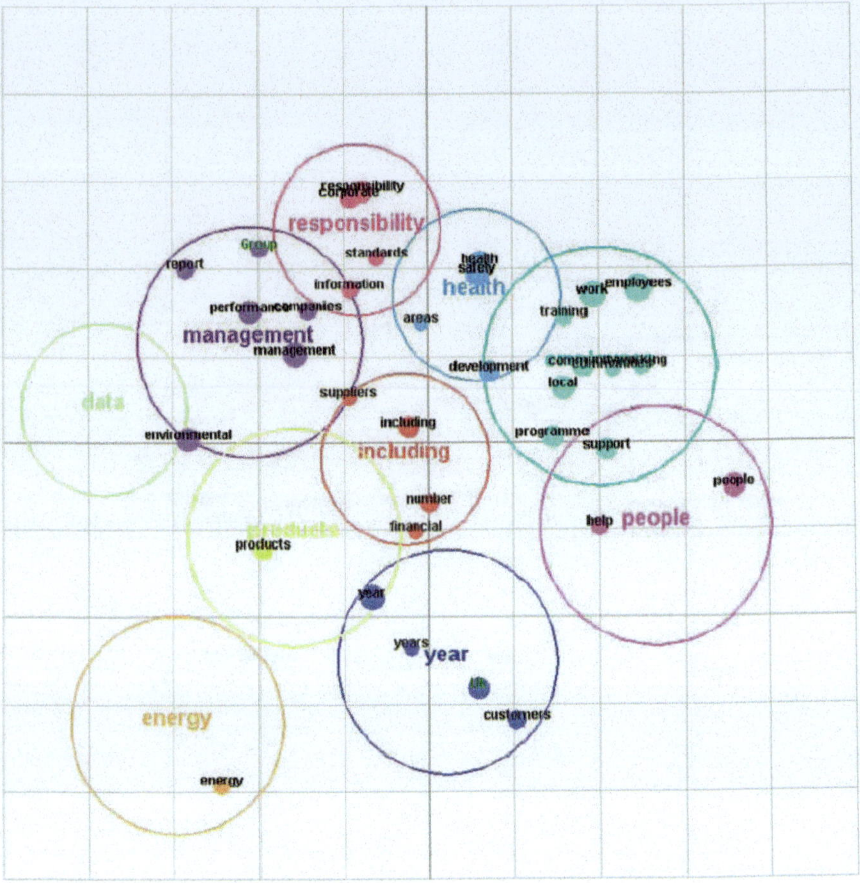

Fig. 12.5 Concept map for UK CSR reports (Chen and Bouvain 2009)

Catholic traditions shaped the French mentality, where property ownership and commerce were considered appropriate only if they are not excessive. Later on, with the arrival of the first Socialist government in 1936, social issues became even more prominent in French political discourse. New legislation on the "*bilan social*" introduced in 1977 was later used in the preparation of the NRE and resulted in a rather limited coverage of social issues, mostly confined to employment relationships, excluding human rights issues. At the moment, companies are faced with an ambiguous situation where they have to comply with French law to include sustainability issues in their general reports complying with NRE and also produce separate CSR reports according to GRI guidelines.

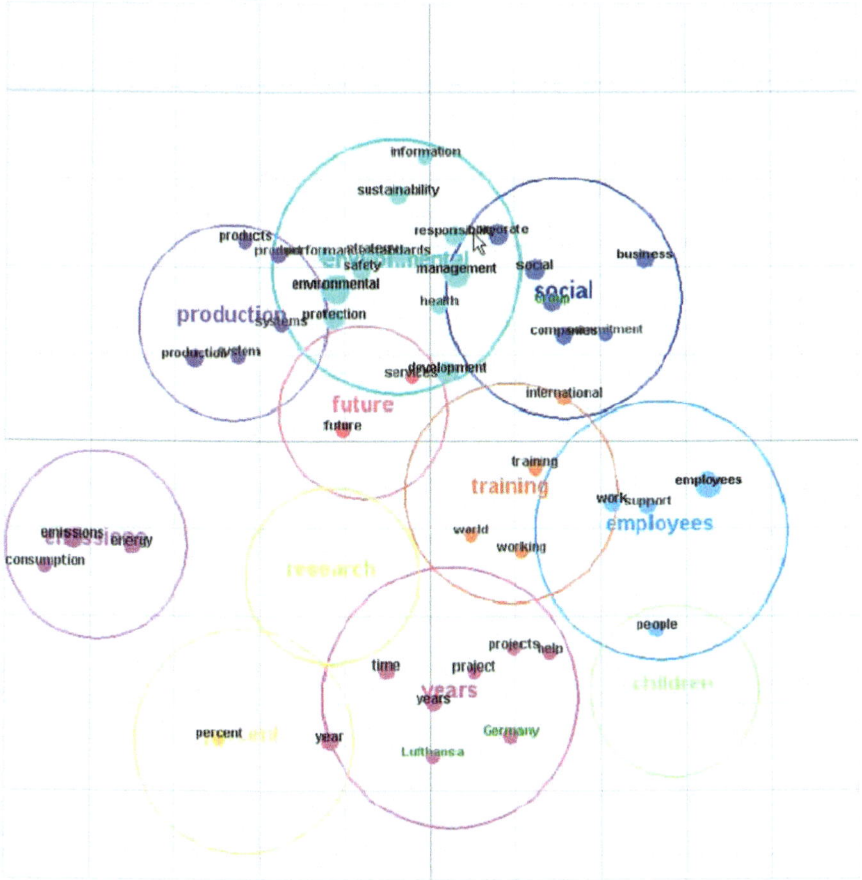

Fig. 12.6 Concept map for German CSR reports (Chen and Bouvain 2009)

The cross-cultural differences in the scope and the emphasis of CSR reports are probably best illustrated by the sophisticated semantic differential analysis carried out by (Chen and Bouvain 2009) with the help of Leximancer Software. Analysing hundreds of CSR reports produced in the USA, UK, Australia and Germany the authors identified six major themes which commonly recurred in the reports: workers, customers, suppliers, community, environment, and society. As Figs. 12.4–12.6 show, the concept maps for each of the chosen countries are quite different. In the US CSR reports, a relatively high importance is placed on community and employee-related issues; in the UK reports, employee and community related issues remain significant, but are related to health and safety issues. German company reports are clearly very distinct from all other countries in the sample. While employees remain central, there is much clear emphasis on environmental and social issues.

References

Angus-Leppan T, Benn S and Young L (2010) A Sensemaking Approach to Trade-Offs and Synergies Between Human and Ecological Elements of Corporate Sustainability *Business Strategy and the Environment* 19.4 230-244

Atkinson G (2000) Measuring Corporate Sustainability *Journal of Environmental Planning and Management* 43.2 235-252

Azapagic A (2003) Systems Approach to Corporate Sustainability: A General Management Framework *Process Safety and Environmental Protection* 81.5 303-316

Blasco M and Zølner M (2010) Corporate Social Responsibility in Mexico and France *Business & Society* 49.2 216-251

Brown HS, de Jong M and Levy DL (2009) Building Institutions Based on Information Disclosure: Lessons from GRI's Sustainability Reporting *Journal of Cleaner Production* 17.6 571-580

Brown HS, de Jong M and Lessidrenska T (2009) The Rise of the Global Reporting Initiative: A Case of Institutional Entrepreneurship *Environmental Politics* 18.2 182-200

Callens I and Tyteca D (1999) Towards Indicators of Sustainable Development for Firms: A Productive Efficiency Perspective *Ecological Economics* 28.1 41-53

Carroll AB (1974) Corporate Social Responsibility: Its Managerial Impact and Implications *Journal of Business Research* 2.1 75-88

Carroll AB (1979) A Three-Dimensional Conceptual Model of Corporate Performance *The Academy of Management Review* 4.4 497-505

Carroll AB (1991) The Pyramid of Corporate Social Responsibility: Toward the Moral Management of Organizational Stakeholders *Business Horizons* 34.4 39-48

Carroll AB (1999) Corporate Social Responsibility: Evolution of a Definitional Construct *Business & Society* 38.3 268-295

Chen S and Bouvain P (2009) Is Corporate Responsibility Converging? A Comparison of Corporate Responsibility Reporting in the USA, UK, Australia, and Germany *Journal of Business Ethics* 87.0 299-317

Clifton D and Amran A (2011) The Stakeholder Approach: A Sustainability Perspective *Journal of Business Ethics* 98.1 121-136

Daub C-H (2007) Assessing the Quality of Sustainability Reporting: An Alternative Methodological Approach *Journal of Cleaner Production* 15.1 75-85

Delbard O (2008) CSR Legislation in France and the European Regulatory Paradox: An Analysis of EU CSR Policy and Sustainability Reporting Practice *Corporate Governance* 8.4 397-405

Duarte F (2010) Working with Corporate Social Responsibility in Brazilian Companies: The Role of Managers Values in the Maintenance of CSR Cultures *Journal of Business Ethics* 96.3 355-368

Dunphy D (2003) *Organizational Change for Corporate Sustainability: A Guide for Leaders and Change Agents of the Future* London/New York, Routledge

Dyllick T (2002) Beyond the Business Case for Corporate Sustainability *Business Strategy and the Environment* 11.2 130-141

Figge F and Hahn T (2004) Sustainable Value Added – Measuring Corporate Contributions to Sustainability Beyond Eco-Efficiency *Ecological Economics* 48.2 173-187

Fukukawa K and Teramoto Y (2009) Understanding Japanese CSR: The Reflections of Managers in the Field of Global Operations *Journal of Business Ethics* 85.0 133-146

Gamerschlag R, Möller K and Verbeeten F (2010) Determinants of Voluntary CSR Disclosure: Empirical Evidence from Germany *Review of Managerial Science* 4 1-30

Hand K and Charity Finance Directors' Group (Great Britain) (2009) *Sustainability in Practice: Monitoring and Reporting* London, Charity Finance Directors' Group

Hedberg C-J and von Malmborg F (2003) The Global Reporting Initiative and Corporate Sustainability Reporting in Swedish Companies *Corporate Social Responsibility and Environmental Management* 10.3 153-164

Henriques A (2004) *The Triple Bottom Line, Does It All Add Up?: Assessing the Sustainability of Business and CSR* London/Sterling, Earthscan

Idowu SO and Towler BA (2004) A Comparative Study of the Content of Corporate Social Responsibility Reports of UK Companies *Management of Environmental Quality* 15.4 420-437

Isaksson R and Steimle U (2009) What Does GRI-Reporting Tell Us About Corporate Sustainability? *The TQM Journal* 21.2 168-181

Line M, Hawley H and Krut R (2002) The Development of Global Environmental and Social Reporting *Corporate Environmental Strategy* 9.1 69-78

Miles M, Munilla L and Darroch J (2009) Sustainable Corporate Entrepreneurship *International Entrepreneurship and Management Journal* 5.1 65-76

Nitkin D and Brooks LJ (1998) Sustainability Auditing and Reporting: The Canadian Experience *Journal of Business Ethics* 17.13 1499-1507

O'Connor M and Spangenberg JH (2008) A Methodology for CSR Reporting: Assuring a Representative Diversity of Indicators Across Stakeholders, Scales, Sites and Performance Issues *Journal of Cleaner Production* 16.13 1399-1415

Steger U (2004) *The Business of Sustainability: Building Industry Cases for Corporate Sustainability* Basingstoke/New York, Palgrave Macmillan

Stubbs W and Cocklin C (2008) Conceptualizing a "Sustainability Business Model" *Organization & Environment* 21.2 103-127

Taneja S, Taneja P and Gupta R (2011) Researches in Corporate Social Responsibility: A Review of Shifting Focus, Paradigms, and Methodologies *Journal of Business Ethics* 101.3 1-22

Tschopp DJ (2005) Corporate Social Responsibility: A Comparison Between the United States and the European Union *Corporate Social Responsibility and Environmental Management* 12.1 55-59

Ulhoi JP (1995) Corporate Environmental and Resource Management: In Search of a New Managerial Paradigm *European Journal of Operational Research* 80.1 2-15

Van Tulder R and Van der Zwart (2006) *International Business-Society Management: Linking Corporate Responsibility and Globalization* London, Routledge

Verbeke A (2009) *International Business Strategy: Rethinking the Foundations of Global Corporate Success* Cambridge, Cambridge University Press

Wang L and Lin L (2007) A Methodological Framework for the Triple Bottom Line Accounting and Management of Industry Enterprises *International Journal of Production Research* 45.5 1063-1088

Welford R (2002) Globalization, Corporate Social Responsibility and Human Rights *Corporate Social Responsibility and Environmental Management* 9.1 1-7

Welford R, Chan C and Man M (2008) Priorities for Corporate Social Responsibility: A Survey of Businesses and Their Stakeholders *Corporate Social Responsibility and Environmental Management* 15.1 52-62

Werther W (2011) *Strategic Corporate Social Responsibility: Stakeholders in a Global Environment* (2nd edition) Los Angeles, Sage

Xinyu X, Yi Z and Longhua X (2010) An Evaluation of Social Responsibility Reporting in China *The China Nonprofit Review* 2.1 101-128

Index

S.E. Shmelev, *Ecological Economics: Sustainability in Practice*,
DOI 10.1007/978-94-007-1972-9,

The manufacturer's authorised representative in the EU is Springer Nature Customer Service Centre GmbH, Europaplatz 3, 69115 Heidelberg, Germany. If you have any concerns regarding our products, please contact ProductSafety@springernature.com

Printed and bound by CPI Group (UK) Ltd, Croydon, CR0 4YY
30/09/2026
02225501-0003